Atlantis Studies in Differential Equations

Volume 3

Series editor

Michel Chipot, Zürich, Switzerland

Aims and Scope of the Series

The “Atlantis Studies in Differential Equations” publishes monographs in the area of differential equations, written by leading experts in the field and useful for both students and researchers. Books with a theoretical nature will be published alongside books emphasizing applications.

For more information on this series and our other book series, please visit our website at: www.atlantis-press.com/publications/books

AMSTERDAM – PARIS – BEIJING
ATLANTIS PRESS
Atlantis Press
29, avenue Laumière
75019 Paris, France

More information about this series at www.atlantis-press.com

Michael I. Gil'

Stability of Neutral Functional Differential Equations

ATLANTIS PRESS

Michael I. Gil'
Department of Mathematics
Ben-Gurion University of the Negev
Beer Sheva
Israel

ISSN 2214-6253 ISSN 2214-6261 (electronic)
ISBN 978-94-6239-090-4 ISBN 978-94-6239-091-1 (eBook)
DOI 10.2991/978-94-6239-091-1

Library of Congress Control Number: 2014950675

Printed on acid-free paper

Preface

1. The suggested book deals with the stability of linear and nonlinear vector neutral type functional differential equations. Equations with neutral type linear parts and nonlinear causal mappings are also considered. Explicit conditions for the exponential, absolute and input-to-state stabilities are derived. Moreover, solution estimates for the considered equations are established. These estimates provide the bounds for regions of attraction of steady states. The main methodology presented in the book is based on a combined usage of the recent norm estimates for matrix-valued functions with the following methods and results: the generalized Bohl-Perron principle, the integral version of the generalized Bohl-Perron principle and the positivity conditions for fundamental solutions to scalar neutral equations. We also apply the so-called generalized norm. A significant part of the book is devoted to the generalized Aizerman problem.
2. Neutral type functional differential equations (NDEs) naturally arise in various applications, such as control systems, mechanics, nuclear reactors, distributed networks, heat flows, neural networks, combustion, interaction of species, microbiology, learning models, epidemiology, physiology, and many others. The theory of functional differential equations has been developed in the works of V. Volterra, A.D. Myshkis, N.N. Krasovskii, B. Razumikhin, N. Minorsky, R. Bellman, A. Halanay, J. Hale and other mathematicians.
   The problem of stability analysis of various neutral type equations continues to attract the attention of many specialists despite its long history. It is still one of the most burning problems because of the absence of its complete solution. The basic method for the stability analysis is the method of the Lyapunov type functionals. By that method many very strong results are obtained. We do not consider the Lyapunov functionals method because several excellent books cover this topic. It should be noted that finding the Lyapunov type functionals for vector neutral type equations is often connected with serious mathematical difficulties, especially in regard to nonautonomous and nonlinear equations. On the contrary, the stability conditions presented in the suggested book are mainly formulated in terms of the determinants and eigenvalues of auxiliary matrices

dependent on a parameter. This fact allows us to apply the well-known results of the theory of matrices to the stability analysis.

3. The aim of the book is to provide new tools for specialists in the stability theory of functional differential equations, control system theory and mechanics. This is the first book that:

   (i) gives a systematic exposition of the approach to stability analysis of vector neutral type functional differential equations which is based on estimates for matrix-valued functions allowing us to investigate various classes of equations from the unified viewpoint;
   (ii) contains a solution of the generalized Aizerman problem for NDEs;
   (iii) presents the generalized Bohl-Perron principle for neutral type systems and its integral version;
   (iv) suggests explicit stability conditions for semilinear equations with linear neutral type parts and nonlinear causal mappings.

   The book is intended not only for specialists in the theory of functional differential equations, but for anyone interested in various applications who has had at least a first year graduate level course in analysis.

   I was very fortunate to have fruitful discussions with the late Professors M.A. Aizerman, M.A. Krasnosel'skii, A.D. Myshkis, A. Pokrovskii and A.A. Voronov, to whom I am very grateful for their interest in my investigations.

4. The book consists of 9 chapters.
   Chapter 1 is of preliminary character. In that chapter we present standard facts, mainly from the theory of Banach and ordered spaces, required in further consideration. In addition, we establish norm estimates for operators of the forms

$$f \to \int_0^\eta dR_0(s)\, f(t-s)$$

and

$$f \to \int_0^\eta d_s R(t,s)\, f(t-s) \quad (t \geq 0),$$

where $R_0(s)$ and $R(t, s)$ are real matrix valued functions having bounded variations in s. These estimates play an essential role in the book.

In Chap. 2 we accumulate norm estimates for matrix-valued functions and bounds for the eigenvalues of matrices. In particular, we suggest estimates for the resolvents, powers of matrices as well as for matrix exponentials. We also present bounds for the spectral radius and investigate perturbations of eigenvalues. The material of this paper is systematically applied in the rest of the book chapters.

Chapter 3 deals with linear autonomous and time-variant vector difference-delay equations with continuous time. Under various assumptions, norm estimates for the Cauchy (resolvent) operators of the considered equations are derived. In the sequel chapters these estimates give us stability conditions for the neutral type functional differential equations.

In Chap. 4 we consider vector linear differential-delay equations (DDEs). We derive estimates for the L^p- and C-norms of the Cauchy operators of autonomous and time-variant differential-delay equations. These estimates enable us to investigate neutral type equations.

Chapter 5 is devoted to vector linear autonomous NDEs. We derive estimates for the L^p- and C-norms of the characteristic matrix functions and fundamental solutions of the considered equations, as well as establish stability conditions for autonomous NDEs.

In Chap. 6 we investigate linear vector time-variant (non-autonomous) neutral type functional differential equations. In particular, we extend the Bohl-Perron principle to a class of neutral type functional differential equations. Namely, it is proved that the homogeneous equation is exponentially stable, provided the corresponding non-homogeneous equation with the zero initial condition and an arbitrary bounded free term has a bounded solution. We also establish the integral version of the generalized Bohl-Perron principle for NDEs, i.e. it is shown that the homogeneous equation is exponentially stable, if the corresponding non-homogeneous equation with the zero initial condition and an arbitrary free term from $L^p(0,\infty) = L^p([0,\infty), \mathbf{C}^n)$, has a solution belonging to $L^p(0,\infty)$. As applications of these principles, the stability conditions for time-variant NDEs close to autonomous systems are derived. In addition, we investigate time-variant systems with small principal operators and obtain stability conditions independent of delay in the non-autonomous case.

Chapter 7 deals with nonlinear vector equations having linear autonomous neutral type parts and nonlinear causal mappings. Explicit conditions for the exponential, absolute and input-to-state stabilities are derived.

Chapter 8 is concerned with scalar nonlinear neutral type functional differential equations with autonomous linear parts. We derive explicit absolute L^2-stability conditions in terms of the norms of the Cauchy operators to the linear parts and the Lipschitz constants of the nonlinearities. In addition, we consider the generalized Aizerman problem.

In Chap. 9 we discuss certain properties of the characteristic values of autonomous vector NDEs. In particular, bounds for characteristic values and perturbations results are derived.

Contents

Symbols

$\|x\|_n$	Euclidean norm of vector x
$\|A\|_n$	Spectral (operator) norm of matrix A
$(.,.)$	Scalar product
$\|A\|$	Matrix whose elements are absolute values of A
A^{-1}	Inverse to A
A^*	Adjoint to A
$\mathbf{C}$	Complex plane
$\mathbf{C}^n$	Complex Euclidean space
$det(A)$	Determinant of A
$I = I_X$	Identity operator in a space X
$I = I_n$	Identity operator in $\mathbf{C}^n$
$\lambda_k(A)$	Eigenvalue of A
$N_p(A)$	Neumann–Schatten norm of A
$N_2(A)$	Hilbert–Schmidt (Frobenius) norm of A
$\mathbf{R}$	Real line
$\mathbf{R}^n$	Real Euclidean space
$R_\lambda(A)$	Resolvent of A
$r_s(A)$	Spectral radius of A
$r_l(A)$	Lower spectral radius of A
$\rho(A, \lambda)$	Distance between λ and the spectrum of A
$s_j(A)$	s-number (singular number) of A
$sv_A(B)$	Spectral variation of B with respect to A
$\sigma(A)$	Spectrum of operator A
$\Sigma(A(.))$	Spectrum of pencil $A(.)$
$Tr\,A = Trace\,A$	Trace of A

Chapter 1
Preliminaries

This chapter is of preliminary character. Here we present standard facts, mainly from the theory of Banach and ordered spaces, required in further consideration. In addition, we establish norm estimates for the operators of the forms

$$f \to \int_0^\eta dR_0(s)f(t-s)$$

and

$$f \to \int_0^\eta d_sR(t,s)f(t-s) \quad (t \geq 0),$$

where $R_0(s)$ is a real matrix valued function defined on a real finite segment $[0, \eta]$, having a bounded variation, and $R(t, s)$ is a real matrix valued function having bounded variation in s.

These estimates play an essential role in the book.

1.1 Banach Spaces and Linear Operators

In Sects. 1.1–1.3 we recall very briefly some basic notions of the theory of Banach and Hilbert spaces. More details can be found in any textbook on Banach and Hilbert spaces (e.g. Ahiezer and Glazman 1981 and Dunford and Schwartz 1966).

Denote the set of complex numbers by **C** and the set of real numbers by **R**.

A linear space X over **C** is called *a (complex) linear normed space* if for any $x \in X$ a non-negative number $\|x\|_X = \|x\|$ is defined, called the norm of x, having the following properties:

1. $\|x\| = 0$ iff $x = 0$,

M.I. Gil', *Stability of Neutral Functional Differential Equations*,
Atlantis Studies in Differential Equations 3, DOI 10.2991/978-94-6239-091-1_1

2. $\|\alpha x\| = |\alpha| \|x\|$,
3. $\|x + y\| \leq \|x\| + \|y\|$ for every $x, y \in X$, $\alpha \in \mathbf{C}$.

A sequence $\{h_n\}_{n=1}^{\infty}$ of elements of X converges *strongly* (in the norm) to $h \in X$ if

$$\lim_{n\to\infty} \|h_n - h\| = 0.$$

A sequence $\{h_n\}$ of elements of X is called the fundamental (Cauchy) one if

$$\|h_n - h_m\| \to 0 \text{ as } m, n \to \infty.$$

If any fundamental sequence converges to an element of X, then X is called *a (complex) Banach space.*

Let in a linear space H over $\mathbf{C}$ for all $x, y \in H$ a number (x, y) be defined, such that

1. $(x, x) > 0$, if $x \neq 0$, and $(x, x) = 0$, if $x = 0$,
2. $(x, y) = \overline{(y, x)}$,
3. $(x_1 + x_2, y) = (x_1, y) + (x_2, y)$ $(x_1, x_2 \in H)$,
4. $(\lambda x, y) = \lambda(x, y)$ $(\lambda \in \mathbf{C})$.

Then $(., .)$ is called the scalar product. Define in H the norm by

$$\|x\| = \sqrt{(x, x)}.$$

If H is a Banach space with respect to this norm, then it is called *a Hilbert space.* The Schwarz inequality

$$|(x, y)| \leq \|x\| \, \|y\|$$

is valid.

If, in an infinite dimensional Hilbert space, there is a countable set whose closure coincides with the space, then that space is said to be *separable*. Any separable Hilbert space H possesses an orthonormal basis. This means that there is a sequence $\{e_k \in H\}_{k=1}^{\infty}$ such that

$$(e_k, e_j) = 0 \text{ if } j \neq k \quad \text{and } (e_k, e_k) = 1 \quad (j, k = 1, 2, \ldots),$$

and any $h \in H$ can be represented as

$$h = \sum_{k=1}^{\infty} c_k e_k$$

with

$$c_k = (h, e_k), \quad k = 1, 2, \ldots.$$

Besides the series strongly converges.

An operator A, acting from a Banach space X into a Banach space Y, is called a linear one if

$$A(\alpha x_1 + \beta x_2) = \alpha A x_1 + \beta A x_2$$

for any $x_1, x_2 \in X$ and $\alpha, \beta \in \mathbf{C}$. If there is a constant a, such that the inequality

$$\|Ah\|_Y \leq a\|h\|_X \text{ for all } h \in X$$

holds, then the operator is said to be bounded. The quantity

$$\|A\| = \|A\|_{X \to Y} := \sup_{h \in X} \frac{\|Ah\|_Y}{\|h\|_X}$$

is called the norm of A. If $X = Y$ we will write $\|A\|_{X \to X} = \|A\|_X$ or simply $\|A\|$.

Under the natural definitions of addition and multiplication by a scalar, and the norm, the set $B(X, Y)$ of all bounded linear operators acting from X into Y becomes a Banach space. If $Y = X$ we will write $B(X, X) = B(X)$. A sequence $\{A_n\}$ of bounded linear operators from $B(X, Y)$ converges *in the uniform operator topology* (in the operator norm) to an operator A if

$$\lim_{n \to \infty} \|A_n - A\|_{X \to Y} = 0.$$

A sequence $\{A_n\}$ of bounded linear operators *converges strongly* to an operator A if the sequence of elements $\{A_n h\}$ strongly converges to Ah for every $h \in X$.

If ϕ is a linear operator, acting from X into $\mathbf{C}$, then it is called a linear functional. It is bounded (continuous) if $\phi(x)$ is defined for any $x \in X$, and there is a constant a such that the inequality

$$|\phi(h)| \leq a\|h\|_X \text{ for all } h \in X$$

holds. The quantity

$$\|\phi\| := \sup_{h \in X} \frac{|\phi(h)|}{\|h\|_X}$$

is called *the norm of the functional* ϕ. All linear bounded functionals on X form a Banach space with that norm. This space is called the space *dual* to X and is denoted by X^*.

In the sequel $I_X = I$ is the identity operator in $X : Ih = h$ for any $h \in X$.

The operator $A^{-1} : Y \to X$ is the inverse one to $A \in B(X, Y)$ if $AA^{-1} = I_Y$ and $A^{-1}A = I_X$.

Let $A \in B(X, Y)$. Consider a linear bounded functional f defined on Y. Then on X the linear bounded functional $g(x) = f(Ax)$ is defined. The operator realizing the relation $f \to g$ is called the operator A^* *dual (adjoint)* to A. By the definition

$$(A^*f)(x) = f(Ax) \ (x \in X).$$

The operator A^* is a bounded linear operator acting from Y^* to X^*. Moreover, the relation $\|A\| = \|A^*\|$ is true.

Theorem 1.1 (Banach-Steinhaus) *Let $\{A_k\}$ be a sequence of linear operators acting from a Banach space X to a Banach space Y. Let for each $h \in X$,*

$$\sup_k \|A_k h\|_Y < \infty.$$

Then the operator norms of $\{A_k\}$ are uniformly bounded. Moreover, if $\{A_n\}$ strongly converges to a (linear) operator A, then

$$\|A\|_{X \to Y} \le \sup_n \|A_n\|_{X \to Y}.$$

For the proof see, for example, Dunford and Schwartz (1966, p. 66).

A point λ of the complex plane is said to be a regular point of an operator A, if the operator $R_\lambda(A) := (A - I\lambda)^{-1}$ (the resolvent) exists and is bounded. The complement of all regular points of A in the complex plane is the *spectrum* of A. The spectrum of A is denoted by $\sigma(A)$.

The quantity

$$r_s(A) = sup_{s \in \sigma(A)} |s|$$

is the *spectral radius* of A. The Gel'fand formula

$$r_s(A) = \lim_{k \to \infty} \sqrt[k]{\|A^k\|}$$

is valid. The limit always exists. Moreover,

$$r_s(A) \le \sqrt[k]{\|A^k\|}$$

for any integer $k \ge 1$. So

$$r_s(A) \le \|A\|.$$

If there is a nontrivial solution e of the equation $Ae = \lambda(A)e$, where $\lambda(A)$ is a number, then this number is called an eigenvalue of operator A, and $e \in H$ is an eigenvector corresponding to $\lambda(A)$. Any eigenvalue is a point of the spectrum. An eigenvalue $\lambda(A)$ has the (algebraic) multiplicity $r \le \infty$ if

$$dim \ (\cup_{k=1}^{\infty} ker \ (A - \lambda(A)I)^k) = r.$$

In the sequel $\lambda_k(A),\ k = 1, 2, \ldots$ are the eigenvalues of A repeated according to their multiplicities.

A vector v satisfying $(A-\lambda(A)I)^n v = 0$ for a natural n, is a root vector of operator A corresponding to $\lambda(A)$.

An operator V is called a quasinilpotent one, if its spectrum consists of zero, only.

On a linear manifold $D(A)$ of a Banach space X let be defined a linear operator A, mapping $D(A)$ into a Banach space Y. Then $D(A)$ is called the domain of A. A linear operator A is called a closed operator, if from $x_n \in X \to x_0$ and $Ax_n \to y_0$ in the norm, it follows that $x_0 \in D(A)$ and $Ax_0 = y_0$.

A bounded linear operator P is called *a projection* if $P^2 = P$.

Now let us consider operators in a Hilbert space H. A bounded linear operator A^* is adjoint to A, if

$$(Af, g) = (f, A^*g) \text{ for every } h, g \in H.$$

A bounded operator A *is a selfadjoint* one, if $A = A^*$. A is *a unitary operator*, if $AA^* = A^*A = I$. A selfadjoint operator A is positive (negative) definite, if

$$(Ah, h) \geq 0 \ \ ((Ah, h) \leq 0) \text{ for every } h \in H.$$

A selfadjoint operator A is strongly positive (strongly negative) definite, if there is a constant $c > 0$, such that

$$(Ah, h) \geq c\ (h, h) \ \ ((Ah, h) \leq -c\ (h, h)) \text{ for every } h \in H.$$

A bounded linear operator satisfying the relation $AA^* = A^*A$ is called *a normal operator*. It is clear that unitary and selfadjoint operators are examples of normal ones.

The spectrum of a selfadjoint operator is real, the spectrum of a unitary operator lies on the unit circle.

Let P be a a projection in a Hilbert space. If, in addition, $P^* = P$, then it is called *an orthogonal projection (an orthoprojection)*.

1.2 Function Spaces

The following spaces are examples of normed spaces. Besides S means finite or infinite real interval. For more details see Dunford and Schwartz (1966, p. 238).

1. The space $C(S)$ consists of all bounded continuous scalar functions on S with the sup-norm

$$\|f\|_{C(S)} = sup_{s\in S}|f(s)|.$$

2. The space $L^p(S)$ is defined for any real number p, $1 \leq p < \infty$. It consists of those measurable scalar functions on S for which the norm

$$\|f\|_{L^p(S)} = [\int_S |f(s)|^p ds]^{1/p}$$

is finite.

3. The space $L^\infty(S)$ consists of all essentially bounded measurable scalar functions on S. The norm is

$$\|f\|_{L^\infty(S)} = ess \sup_{s \in S} |f(s)|.$$

4. Note that the Hilbert space has been defined by a set of abstract axioms. It is noteworthy that some of the concrete spaces defined above satisfy these axioms, and hence are special cases of abstract Hilbert space. Thus, for instance, the n-dimensional space $\mathbf{C}^n$ is a Hilbert space, if the inner product $(x, y)_{C^n}$ of two elements

$$x = \{x_1, \ldots, x_n\} \text{ and } y = \{y_1, \ldots, y_n\}$$

is defined by the formula

$$(x, y)_{C^n} = \sum_{k=1}^{n} x_k \overline{y}_k.$$

The (Euclidean) norm is $\|x\|_n = \sqrt{(x, x)_{C^n}}$

5. Also the complex space $L^2(S)$ is a Hilbert space with the scalar product

$$(f, g)_{L^2(S)} = \int_S f(s)\overline{g}(s) ds.$$

6. The space $M(S)$ consists of all bounded measurable scalar functions on S with the sup-norm. It is not a Banach space.
7. The space $C^1(S)$ consists of all bounded continuously differentiable scalar functions on S. The norm is

$$\|f\|_{C^1(S)} = sup_{s \in S} |f(s)| + sup_{s \in S} |f'(s)|.$$

8. The space $C(S, \mathbf{C}^n)$ consists of all bounded continuous on S functions with values in $\mathbf{C}^n$. The norm is

$$\|f\|_{C(S)} = sup_{s \in S} \|f(s)\|_n.$$

9. The space $L^p(S, \mathbf{C}^n)$ is defined for any real number p, $1 \le p < \infty$, and consists of those measurable functions defined on S with values in $\mathbf{C}^n$, for which the norm

$$\|f\|_{L^p(S,\mathbf{C}^n)} = [\int_S \|f(s)\|_n^p ds]^{1/p}$$

is finite.

10. The space $L^\infty(S, \mathbf{C}^n)$ consists of all essentially bounded measurable functions defined on S with values in $\mathbf{C}^n$. The norm is

$$\|f\|_{L^\infty(S,\mathbf{C}^n)} = ess \sup_{s\in S} \|f(s)\|_n.$$

11. The complex space $L^2(S, \mathbf{C}^n)$ is a Hilbert space with the scalar product

$$(f, g)_{L^2(S,\mathbf{C}^n)} = \int_S (f(s), g(s))_{C^n} ds.$$

12. The space $M(S, \mathbf{C}^n)$ consists of all bounded measurable functions f on S with values in $\mathbf{C}^n$ and the sup-norm. It is not a Banach space.
13. The space $C^1(S, \mathbf{C}^n)$ consists of all bounded continuously differentiable functions on S with values in $\mathbf{C}^n$. The norm is

$$\|f\|_{C^1(S,\mathbf{C}^n)} = \sup_{s\in S} \|f(s)\|_n + \sup_{s\in S} \|f'(s)\|_n.$$

1.3 Some Theorems From Functional Analysis

Let X and Y be Banach spaces. A function $f : X \to Y$ is said to be continuous if for any $\varepsilon > 0$, there is a $\delta > 0$, such that $\|x - y\|_X \leq \delta$ implies $\|f(x) - f(y)\|_Y \leq \varepsilon$.

Theorem 1.2 (The Urysohn theorem) *Let A and B be disjoint closed sets in a Banach space X. Then there is a continuous function f defined on X such that*

$$0 \leq f(x) \leq 1; \; f(x) = 1 \quad \text{for all } x \in A \quad \text{and } f(x) = 0 \quad \text{for all } x \in B.$$

For the proof see, for instance Dunford and Schwartz (1966, p. 15).

Theorem 1.3 (The Closed Graph theorem) *A closed linear map defined on the all of a Banach space, and with values in an Banach space, is continuous.*

For the proof see Dunford and Schwartz (1966, p. 57).

Let Ω_1 and Ω_2 be two measurable sets.

Theorem 1.4 (The Riesz-Thorin theorem) *Assume T is a bounded linear operator from $L^p(\Omega_1)$ to $L^p(\Omega_2)$ and at the same time from $L^q(\Omega_1)$ to $L^q(\Omega_2)$ $(1 \leq p, q \leq \infty)$. Then it is also a bounded operator from $L^r(\Omega_1)$ to $L^r(\Omega_2)$ for any r between p and q. In addition the following inequality for the norms holds:*

$$\|T\|_{L^r(\Omega_1)\to L^r(\Omega_2)} \le \max\{\|T\|_{L^p(\Omega_1)\to L^p(\Omega_2)}, \|T\|_{L^q(\Omega_1)\to L^q(\Omega_2)}\}.$$

For the details see the book Krein (1972, Sect. III.4). For the proof (in a more general situation) see Dunford and Schwartz (1966, Theorem VI.10.11, p. 527).

Theorem 1.5 *Let $f \in L^1(\Omega)$ be a fixed integrable function and let T be the operator of convolution with f, i.e., for each function $g \in L^p(\Omega)$ $(p \ge 1)$ we have*

$$(Tg)(t) = \int\limits_{\Omega} f(t-s)g(s)ds.$$

Then

$$\|Tg\|_{L^p(\Omega)} \le \|f\|_{L^1(\Omega)}\|g\|_{L^p(\Omega)}.$$

For the proof see Dunford and Schwartz (1966, p. 528).

1.4 Ordered Banach Spaces

Following the book (Meyer-Nieberg 1991), let us introduce an inequality relation for normed spaces which can be used analogously to the inequality relation for real numbers.

A non-empty set M with a relation $\le$ is said to be an ordered set, whenever the following conditions are satisfied.

$$(i)\ x \le x \text{ for every } x \in M,$$

$$(ii)\ x \le y \text{ and } y \le x \text{ implies that } x = y \text{ and}$$

$$(iii)\ x \le y \text{ and } y \le z \text{ implies that } x \le z.$$

If, in addition, for any two elements $x, y \in M$ either $x \le y$ or $y \le x$, then M is called a totally ordered set. Let A be a subset of an ordered set M. Then $x \in M$ is called an upper bound of A, if $y \le x$ for every $y \in A$. $z \in M$ is called a lower bound of A, if $y \ge z$ for all $y \in A$. Moreover, if there is an upper bound of A, then A is said to be bounded from above. If there is a lower bound of A, then A is called bounded from below. If A is bounded from above and from below, then we will briefly say that A is order bounded. Denote

$$[x, y] = \{z \in M : \ x \le z \le y\}.$$

That is, $[x, y]$ is an order interval.

An ordered set $(M, \leq)$ is called *a lattice*, if any two elements $x, y \in M$ have a least upper bound denoted by $sup(x, y)$ and a greatest lower bound denoted by $inf(x, y)$. A subset A is order bounded, if it is contained in some order interval.

Definition 1.1 A real vector space E which is also an ordered set is called an ordered vector space, if the order and the vector space structure are compatible in the following sense: if $x, y \in E$, such that $x \leq y$, then $x + z \leq y + z$ for all $z \in E$ and $ax \leq ay$ for any positive number a. If, in addition, $(E, \leq)$ is a lattice, then E is called a Riesz space (or a vector lattice).

Let E be a Riesz space. The *positive cone* E_+ of E consists of all $x \in E$, such that $x \geq 0$. For every $x \in E$ let

$$x^+ = \sup\ (x, 0),\ \ x^- = \inf\ (-x, 0),\ \ |x| = \sup\ (x, -x)$$

be the *positive part, the negative part and the absolute value of* x, respectively.

Example 1.1 Let $E = \mathbf{R}^n$ and

$$R^n_+ = \{(x_1, \ldots, x_n) \in \mathbf{R}^n : x_k \geq 0 \quad \text{for all } k\}.$$

Then R^n_+ is a positive cone and for $x = (x_1, \ldots, x_n), y = (y_1, \ldots, y_n) \in \mathbf{R}^n$, we have

$$x \leq y \quad \text{iff } x_k \leq y_k$$

and

$$|x| = (|x_1|, \ldots, |x_n|).$$

Example 1.2 Let X be a non-empty set and let $B(X)$ be the collection of all bounded real valued functions defined on X. It is a simple and well-known fact that $B(X)$ is a vector space ordered by the positive cone

$$B(X)_+ = \{f \in B(X) : f(t) \geq 0 \text{ for all } t \in X\}.$$

Thus $f \geq g$ holds, if and only if $f - g \in B(X)_+$. Obviously, the function $h_1 = \sup\ (f, g)$ is defined by

$$h_1(t) = max\ \{f(t), g(t)\}$$

and the function $h_2 = \inf\ (f, g)$ is defined by

$$h_2(t) = min\ \{f(t), g(t)\}$$

for every $t \in X$ and $f, g \in B(X)$. This shows that $B(X)$ is a Riesz space and the absolute value of f is $|f(t)|$.

Definition 1.2 Let E be a Riesz space furnished with a norm $\|.\|$, satisfying $\|x\| \leq \|y\|$ whenever $|x| \leq |y|$. In addition, let the space E be complete with respect to that norm. Then E is called a Banach lattice.

The norm $\|.\|$ in a Banach lattice E is said to be *order continuous*, if

$$\inf\{\|x\| : x \in A\} = 0$$

for any down directed set $A \subset E$, such that $\inf\{x \in A\} = 0$, cf. Meyer-Nieberg (1991, p. 86).

The real spaces $C(K)$, $L^p(K)$ $(K \subseteq \mathbf{R}^n)$ and l^p $(p \geq 1)$ are examples of Banach lattices.

A bounded linear operator T in E is called *a positive one*, if from $x \geq 0$ it follows that $Tx \geq 0$.

1.5 The Abstract Gronwall Lemma

In this section E is a Banach lattice with the positive cone E_+.

Lemma 1.1 (The abstract Gronwall lemma) *Let T be a bounded linear positive operator acting in E and having the spectral radius*

$$r_s(T) < 1.$$

Let $x, f \in E_+$. Then the inequality

$$x \leq f + Tx$$

implies $x \leq y$ where y is a solution of the equation

$$y = f + Ty.$$

Proof Let $Bx = f + Tx$. Then $x \leq Bx$ implies $x \leq Bx \leq B^2x \leq \cdots \leq B^m x$. This gives

$$x \leq B^m x = \sum_{k=1}^{m-1} T^k f + T^m x \to (I - T)^{-1} f = y \text{ as } m \to \infty.$$

Since $r_s(T) < 1$, the Neumann series converges. □

We will say that $F : E \to E$ is *a non-decreasing mapping* if $v \leq w$ $(v, w \in E)$ implies $F(v) \leq F(w)$.

Lemma 1.2 *Let $F : E \to E$ be a non-decreasing mapping, and $F(0) = 0$. In addition, let there be a positive linear operator T in E, such that the conditions*

$$|F(v) - F(w)| \le T|v - w| \quad (v, w \in E) \tag{5.1}$$

and $r_s(T) < 1$ hold. Then the inequality

$$x \le F(x) + f \quad (x, f \in E_+)$$

implies that $x \le y$ where y is a solution of the equation

$$y = F(y) + f.$$

Moreover, the inequality

$$z \ge F(z) + f \quad (z, f \in E_+)$$

implies that $z \ge y$.

Proof We have $x = F(x) + h$ with an $h < f$. Thanks to (5.1) and the condition $r_s(T) < 1$, the mappings $F_f := F + f$ and $F_h := F + h$ have the following properties: F_f^m and F_h^m are contracting for some integer m. So thanks to the generalized contraction mapping theorem (Krasnosel'skii and Zabreiko 1984), $F_f^k(f) \to x$, $F_h^k(f) \to y$ as $k \to \infty$. Moreover, $F_f^k(f) \ge F_h^k(f)$ for all $k = 1, 2, \ldots$, since F is non-decreasing and $h \le f$. This proves the inequality $x \ge y$. Similarly the inequality $x \le z$ can be proved. □

1.6 Integral Inequalities

Let $C(J, \mathbf{R}^n)$ be a space of real vector valued functions defined, bounded and continuous on a finite or infinite interval J. The inequalities are understood in the coordinate-wise sense.

To receive various solution estimates, we essentially use the following lemma.

Lemma 1.3 *Let $\hat{K}(t, s)$ be a matrix kernel with non-negative entries, such that the integral operator*

$$(Kx)(t) = \int_J \hat{K}(t, s)x(s)ds$$

maps $C(J, \mathbf{R}^n)$ into itself and has the spectral radius $r_s(K) < 1$. Then for any non-negative continuous vector function $v(t)$ satisfying the inequality

$$v(t) \le \int_J \hat{K}(t, s)v(s)ds + f(t)$$

where f is a non-negative continuous on J vector function, the inequality $v(t) \leq u(t)$ $(t \in J)$ is valid, where $u(t)$ is a solution of the equation

$$u(t) = \int_J \hat{K}(t,s)u(s)ds + f(t).$$

Similarly, the inequality

$$v(t) \geq \int_J \hat{K}(t,s)v(s)ds + f(t)$$

implies $v(t) \geq u(t)$ $(t \in J)$.

Proof The lemma is a particular case of the abstract Gronwall lemma. □

If $J = [a,b]$ is an arbitrary finite interval and

$$(Kx)(t) = \int_a^t \hat{K}(t,s)x(s)ds \ (t \leq b),$$

and the condition

$$\sup_{t \in [a,b]} \int_a^t \|\hat{K}(t,s)\|ds < \infty$$

is fulfilled with an arbitrary matrix norm, then it is simply to show that $r_s(K) = 0$.

The same equality for the spectral radius is true, if

$$(Kx)(t) = \int_t^b \hat{K}(t,s)x(s)ds \ (t \geq a),$$

provided

$$\sup_{t \in [a,b]} \int_t^b \|\hat{K}(t,s)\|ds < \infty.$$

Hence as a particular case of the previous lemma we arrive at

Lemma 1.4 (The Gronwall lemma) *Let $a(t)$ $(t \geq 0)$ be a non-negative bounded measurable scalar function. Then for any non-negative continuous scalar function $v(t)$ satisfying the inequality*

$$v(t) \leq \int_0^t a(s)v(s)ds + c \quad (t \geq 0;\ c = const \geq 0)$$

we have

$$v(t) \leq ce^{\int_0^t a(s)v(s)ds}.$$

Similarly, the inequality

$$v(t) \geq \int_0^t a(s)v(s)ds + c \quad (t \geq 0).$$

implies

$$v(t) \geq ce^{\int_0^t a(s)v(s)ds}.$$

1.7 The Kantorovich Generalized Norm

In this section nonlinear equations are considered in a space furnished with a vector (generalized) norm introduced by Kantorovich, cf. (Vulikh 1967, p. 334). Note that the vector norm enables us to use information about equations more completely than a usual (number) norm.

Throughout this section E is a Banach lattice with a positive cone E_+ and a norm $\|.\|_E$.

Let X be an arbitrary set. Assume that in X a vector metric $M(.,.)$ is defined. That is, $M(.,.)$ maps $X \times X$ into E_+ with the usual properties: for all $x, y, z \in X$

$$a)\ M(x, y) = 0 \text{ iff } x = y;\ \ b)\ M(x, y) = M(y, x)$$

and

$$c)\ M(x, y) \leq M(x, z) + M(y, z).$$

Clearly, X is a metric space with the metric $m(x, y) = \|M(x, y)\|_E$. That is, a sequence $\{x_k \in X\}$ converges to x in the metric $m(.,.)$ iff $M(x_k, x) \to 0$ as $k \to \infty$.

Lemma 1.5 *Let X be a space with a vector metric $M(.,.): X \times X \to E_+$, and $F(x)$ map a closed set $\Phi \subseteq X$ into itself with the property*

$$M(F(x), F(y)) \leq QM(x, y) \quad (x, y \in \Phi), \tag{7.1}$$

where Q is a positive operator in E whose spectral radius $r_s(Q)$ is less than one: $r_s(Q) < 1$. Then, if X is complete in generalized metric $M(.,.)$ (or, equivalently, in metric $m(.,.)$), F has a unique fixed point $\overline{x} \in \Phi$. Moreover, that point can be found by the method of successive approximations.

Proof Following the usual proof of the contracting mapping theorem we take an arbitrary $x_0 \in \Phi$ and define the successive approximations by the equality

$$x_k = F(x_{k-1}) \ \ (k = 1, 2, \ldots).$$

Hence,

$$M(x_{k+1}, x_k) = M(F(x_k), F(x_{k-1})) \le QM(x_k, x_{k-1}) \le \cdots \le Q^k M(x_1, x_0).$$

For $m > k$ we thus get

$$M(x_m, x_k) \le M(x_m, x_{m-1}) + M(x_{m-1}, x_k) \le \cdots \le \sum_{j=k}^{m-1} M(x_{j+1}, x_j)$$
$$\le \sum_{j=k}^{m-1} Q^j M(x_1, x_0).$$

Inasmuch as $r_s(Q) < 1$,

$$M(x_m, x_k) \le Q^k (I - Q)^{-1} M(x_1, x_0) \to 0 \ (k \to \infty).$$

Recall that I is the unit operator in a corresponding space. Consequently, points x_k converge in the metric $M(.,.)$ to an element $\overline{x} \in \Phi$. Since $\lim_{k\to\infty} F(x_k) = F(\overline{x})$, $\overline{x}$ is the fixed point due to (7.1). Thus, the existence is proved.

To prove the uniqueness let us assume that $y \ne \overline{x}$ is a fixed point of F as well. Then by (7.1) $M(\overline{x}, y) = M(F(\overline{x}), F(y)) \le QM(\overline{x}, y)$. Or

$$(I - Q)^{-1} M(\overline{x}, y) \le 0.$$

But $I - Q$ is positively invertible, because $r_s(Q) < 1$. In this way, $M(\overline{x}, y) \le 0$. This proves the result. □

Now let X be a linear space with a vector (generalized) norm $M(.)$. That is, $M(.)$ maps X into E_+ and is subject to the usual axioms: for all $x, y \in X$

$$M(x) > 0 \text{ if } x \ne 0; \ \ M(\lambda x) = |\lambda| M(x) \ \ (\lambda \in \mathbf{C}); \ \ M(x + y) \le M(x) + M(y).$$

Following (Vulikh 1967), we shall call *E a norming lattice, and X a lattice-normed space*. Clearly, X with a generalized (vector) norm $M(.) : X \to E_+$ is a normed space with the norm

$$\|h\|_X = \|M(h)\|_E \ (h \in X). \tag{7.2}$$

Now the previous lemma implies

Corollary 1.1 *Let X be a space with a generalized norm $M(.) : X \to E_+$ and $F(x)$ map a closed set $\Phi \subseteq X$ into itself with the property*

$$M(F(x) - F(y)) \leq QM(x - y) \ \ (x, y \in \Phi),$$

where Q is a positive operator in E with $r_s(Q) < 1$. Then, if X is complete in the norm defined by (7.2), F has a unique fixed point $\bar{x} \in \Phi$. Moreover, that point can be found by the method of successive approximations.

1.8 Autonomous Difference Operators

Let $[a, b]$ be a finite real segment. A scalar function $g : [a, b] \to \mathbf{R}$ is *a function of bounded variation* if

$$var\ (g) = var_{t\in[a,b]}\, g(t) := \sup_P \sum_{i=0}^{n-1} |g(t_{i+1}) - g(t_i)| < \infty,$$

where the supremum is taken over the set of all partitions P of the segment $[a, b]$.

Any function of bounded variation $g : [a, b] \to \mathbf{R}$ is a difference of bounded nondecreasing functions g^+ and g^-: $g = g^+ - g^-$. If g is differentiable and its derivative is integrable then its variation satisfies

$$var\ (g) \leq \int_a^b |g'(s)|ds.$$

For more details see Dunford and Schwartz (1966, p. 140). For a continuous function $w(s)$ we write

$$\int_a^b w(s)|dg(s)| = \int_a^b w(s)d(g^+(s) + g^-(s)).$$

So

$$var\ (g) = \int_a^b |dg(s)|.$$

Let $\|x\|_n$ be the Euclidean norm of a vector x and $\|A\|_n$ be the spectral norm (the operator norm with respect to the vector Euclidean norm) of a matrix A. Recall that spaces $C([a, b], \mathbf{C}^n)$ and $L^p([a, b], \mathbf{C}^n)$ are defined in Sect. 1.2.

Furthermore, let $R_0(s) = (r_{ij}(s))_{i,j=1}^n$ be a real matrix valued function defined on a real finite segment $[0, \eta]$, whose entries have bounded variations

$$var(r_{ij}) = var_{s\in[0,\eta]}\, r_{ij}(s).$$

It is assumed that R_0 has a finite number of jumps.

Introduce the matrix

$$Var(R_0) = (var(r_{ij}))_{i,j=1}^n$$

and put

$$var\ (R_0) := \|Var(R_0)\|_n.$$

So $var\ (R_0)$ is the spectral (operator) norm of matrix $Var\ (R_0)$.

For a positive constant $T \le \infty$ and a vector valued function $f \in C([-\eta, T], \mathbf{C}^n)$ or $f \in L^p([-\eta, T], \mathbf{C}^n)$ $(1 \le p \le \infty)$, put

$$E_0 f(t) = \int_0^\eta dR_0(s) f(t-s) \quad (0 \le t \le T).$$

Throughout the book in the case $f \in C([-\eta, T], \mathbf{C}^n)$ this integral for each t is understood as the Lebesgues-Stieljes integral. In the case of non-continuous $f \in L^p([-\eta, T], \mathbf{C}^n)$ operator E_0 is defined as the following: we prove that under consideration E_0 is bounded in the norm of L^p $(p \ge 1)$ on the set of continuous functions. Since the set of continuous functions is dense in L^p we take a sequence of continuous functions f_n converging to f in the norm of L^p and put $E_0 f = \lim_{n\to\infty} E_0 f_n$.

Lemma 1.6 *Suppose all the entries r_{jk} of the matrix function R_0 defined on $[0, \eta]$ have bounded variations and finite numbers of jumps. Then for all $T > 0$, the following inequalities are valid:*

$$\|E_0 f\|_{C([0,T],\mathbf{C}^n)} \le \sqrt{n}\ var\ (R_0)\|f\|_{C([-\eta,T],\mathbf{C}^n)} \quad (f \in C([-\eta, T], \mathbf{C}^n)), \tag{8.1}$$

$$\|E_0 f\|_{L^\infty([0,T],\mathbf{C}^n)} \le \sqrt{n}\ var\ (R_0)\|f\|_{L^\infty([-\eta,T],\mathbf{C}^n)} \quad (f \in L^\infty([-\eta, T], \mathbf{C}^n)), \tag{8.2}$$

$$\|E_0 f\|_{L^2([0,T],\mathbf{C}^n)} \le var\ (R_0)\|f\|_{L^2([-\eta,T],\mathbf{C}^n)} \quad (f \in L^2([-\eta, T], \mathbf{C}^n)) \tag{8.3}$$

and

$$\|E_0 f\|_{L^1([0,T],\mathbf{C}^n)} \le \varsigma(R_0)\|f\|_{L^1([-\eta,T],\mathbf{C}^n)} \quad (f \in L^1([-\eta, T], \mathbf{C}^n)), \tag{8.4}$$

where

$$\varsigma(R_0) := \sum_{j=1}^{n} \sqrt{\sum_{k=1}^{n} (var(r_{jk}))^2}.$$

Proof First, let $f(t) = (f_k(t))_{k=1}^n \in C([-\eta, T], \mathbf{C}^n)$. For each coordinate $(E_0 f)_j(t)$ of $E_0 f(t)$, $t \in [0, T]$, we have

$$|(E_0 f)_j(t)| = |\int_0^\eta \sum_{k=1}^n f_k(t-s) dr_{jk}(s)|$$
$$\leq \sum_{k=1}^n \int_0^\eta |dr_{jk}| \max_{0 \leq s \leq \eta} |f_k(t-s)|$$
$$= \sum_{k=1}^n var(r_{jk}) \max_{0 \leq s \leq \eta} |f_k(t-s)|.$$

Hence,

$$\sum_{j=1}^n |(E_0 f)_j(t)|^2 \leq \sum_{j=1}^n \left(\sum_{k=1}^n var(r_{jk}) \|f_k\|_{C(-\eta,T)} \right)^2$$

$$= \|Var\ (R_0)\ \nu_C\|_n^2 \leq (var(R_0) \|\nu_C\|_n)^2,$$

where ν_C is a vector whose coordinates are

$$\|f_k\|_{C(-\eta,T)},\ k = 1, \ldots, n.$$

Here $\|.\|_{C(-\eta,T)} = \|.\|_{C([-\eta,T],\mathbf{C})}$. But

$$\|\nu_C\|_n^2 = \sum_{k=1}^n \|f_k\|_{C(-\eta,T)}^2 \leq n \max_k \|f_k\|_{C(-\eta,T)}^2$$
$$\leq n \sup_t \sum_{k=1}^n \|f_k(t)\|_n^2 = n \|f\|_{C([-\eta,T],\mathbf{C}^n)}^2.$$

So

$$\|E_0 f\|_{C([0,T],\mathbf{C}^n)} \leq \sqrt{n}\ var\ (R_0) \|f\|_{C([-\eta,T],\mathbf{C}^n)}$$

and thus inequality (8.1) is proved.

In the case of the space L^∞ by inequality (8.1) we have

$$\|E_0 f\|_{L^\infty([0,T],\mathbf{C}^n)} \le \sqrt{n}\; var\;(R_0)\|f\|_{L^\infty([-\eta,T],\mathbf{C}^n)}$$

for a continuous function f. But the set of continuous functions is dense in L^∞. So the previous inequality is valid on the whole space L^∞. This proves (8.2).

Now consider the norm in space L^2. We have

$$\int_0^T |(E_0 f)_j(t)|^2 dt \le \int_0^T \Big(\sum_{k=1}^n \int_0^\eta |f_k(t-s)||dr_{jk}(s)|\Big)^2 dt$$

$$= \int_0^\eta \int_0^\eta \sum_{i=1}^n \sum_{k=1}^n |dr_{jk}(s)||dr_{ji}(s_1)| \int_0^T |f_k(t-s) f_i(t-s_1)| dt.$$

By the Schwarz inequality

$$\left(\int_0^T |f_k(t-s) f_i(t-s_1)| dt\right)^2 \le \int_0^T |f_k(t-s)|^2 dt \int_0^T |f_i(t-s_1)|^2 dt$$
$$\le \int_{-\eta}^T |f_k(t)|^2 dt \int_{-\eta}^T |f_i(t)|^2 dt.$$

Thus

$$\int_0^T |(E_0 f)_j(t)|^2 dt \le \sum_{i=1}^n \sum_{k=1}^n var(r_{jk}) var(r_{ji}) \|f_k\|_{L^2(-\eta,T)} \|f_i\|_{L^2(-\eta,T)}$$
$$= \Big(\sum_{k=1}^n var(r_{jk}) \|f_k\|_{L^2(-\eta,T)}\Big)^2$$

where

$$\|f_k\|_{L^2(-\eta,T)} = \|f_k\|_{L^2([-\eta,T],\mathbf{C})}.$$

Therefore

$$\sum_{j=1}^n \int_0^T |(E_0 f)_j(t)|^2 dt \le \sum_{j=1}^n \Big(\sum_{k=1}^n var(r_{jk}) \|f_k\|_{L^2(-\eta,T)}\Big)^2$$

$$= \|Var\;(R_0)\; v_2\|_n^2 \le (var(R_0)\|v_2\|_n)^2$$

where v_2 is the vector with the coordinates $\|f_k\|_{L^2(-\eta,T)}$. But

$$\|v_2\|_n = \|f\|_{L^2([-\eta,T],\mathbf{C}^n)}.$$

So (8.3) is also proved.

Similarly, for an

$$f(t) = (f_k(t))_{k=1}^n \in L^1([-\eta, T], \mathbf{C}^n)$$

we obtain

$$\int_0^T |(E_0 f)_j(t)|dt \le \sum_{k=1}^n \int_0^\eta \int_0^T |f_k(t-s)|dt|dr_{jk}(s)|$$
$$\le \sum_{k=1}^n var(r_{jk}) \int_{-\eta}^T |f_k(t)|dt.$$

So

$$\|E_0 f\|_{L^1([0,T],\mathbf{C}^n)} = \int_0^T \sqrt{\sum_{j=1}^n |(E_0 f)_j(t)|^2}dt \le \int_0^T \sum_{j=1}^n |(E_0 f)_j(t)|dt$$
$$\le \sum_{j=1}^n \int_{-\eta}^T \sum_{k=1}^n var(r_{jk})|f_k(t)|dt.$$

Consequently, by the Schwarz inequality

$$\|E_0 f\|_{L^1([0,T],\mathbf{C}^n)}$$
$$\le \sum_{j=1}^n \int_{-\eta}^T \sqrt{\sum_{k=1}^n (var(r_{jk}))^2 \sum_{k=1}^n |f_k(t)|^2}\ dt.$$

Hence (8.4) follows. As claimed. □

Take into account that

$$\sum_{j=1}^n \sqrt{b_j} \ge \sqrt{\sum_{j=1}^n b_j} \tag{8.5}$$

for positive numbers b_j. Hence,

$$\sum_{j=1}^{n}\sqrt{\sum_{k=1}^{n}(var(r_{jk}))^2} \geq \sqrt{\sum_{j=1}^{n}\sum_{k=1}^{n}(var(r_{jk}))^2}.$$

That is,

$$\varsigma(R_0) \geq var\ (R_0).$$

Now the Riesz-Thorin theorem (see Sect. 1.3) and previous lemma imply the following result.

Corollary 1.2 *Suppose all the entries r_{jk} of the matrix function R_0 defined on $[0,\eta]$ have bounded variations. Then for all $T > 0$, the inequalities*

$$\|E_0 f\|_{L^p([0,T],\mathbf{C}^n)} \leq \sqrt{n}\ var\ (R_0)\|f\|_{L^p([-\eta,T],\mathbf{C}^n)} \ \text{ if } p \geq 2,$$

and

$$\|E_0 f\|_{L^p([0,T],\mathbf{C}^n)} \leq \varsigma(R_0)\|f\|_{L^p([-\eta,T],\mathbf{C}^n)}, \ \text{ if } 1 \leq p < 2 \ (f \in L^p([-\eta,T],\mathbf{C}^n))$$

are valid.

Furthermore, for a continuous scalar function $w(s)$ denote

$$Vd(R_0,w) := \|(\int_0^\eta |w(\tau)||dr_{jk}|)_{j,k=1}^n\|_n.$$

So $Vd\ (R_0,w)$ is the spectral norm of the matrix whose entries are $\int_0^\eta |w(\tau)||dr_{jk}|$.

Lemma 1.7 *Let $w(s)$ be a continuous scalar function defined on $[0,\eta]$. Then*

$$\left\|\int_0^\eta w(\tau)dR(\tau)f(t-\tau)\right\|_{L^2(0,T)} \leq Vd\ (R_0,w)\|f\|_{L^2(-\eta,T)} \ \ (T>0;\ f \in L^2(-\eta,T)). \tag{8.6}$$

Proof First, let $f(t) = (f_k(t))_{k=1}^n$ be continuous. Put

$$E_1 f(t) = \int_0^\eta w(\tau)dR(\tau)f(t-\tau)\ \ (f(t) = (f_k(t))).$$

Then we obtain

$$\|E_1 f(t)\|^2_{L^2(0,T)} = \sum_{j=1}^n \int_0^T |(E_1 f)_j(t)|^2 dt,$$

where $(E_1 f)_j(t)$ denotes the coordinate of $(E_1 f)(t)$. But

$$
\begin{aligned}
|(E_1 f)_j(t)|^2 &= |\sum_{k=1}^{n} \int_0^{\eta} w(s) f_k(t-s)\, dr_{jk}(s)|^2 \\
&\le \left(\sum_{k=1}^{n} \int_0^{\eta} |w(s)||f_k(t-s)||dr_{jk}(s)| \right)^2 \\
&= \sum_{k=1}^{n} \int_0^{\eta} |w(s)||f_k(t-s)||dr_{jk}(s)| \sum_{i=1}^{n} \int_0^{\eta} |w(s_1)||f_i(t-s_1)||dr_{ji}(s_1)|.
\end{aligned}
$$

Hence,

$$
\int_0^T |(E_1 f)_j(t)|^2 dt \le \int_0^T \int_0^{\eta} \int_0^{\eta} \sum_{i=1}^{n} \sum_{k=1}^{n} |w(s)||dr_{jk}(s)|w(s_1)|dr_{ji}(s_1)||f_k(t-s) f_i(t-s_1)|\, dt.
$$

Thus

$$
\int_0^T |(E_1 f)_j(t)|^2 dt \le \int_0^{\eta} \int_0^{\eta} |w(s)||w(s_1)| \sum_{i=1}^{n} \sum_{k=1}^{n} \int_0^T |f_k(t-s) f_i(t-s_1)|\, dt |dr_{jk}(s)||dr_{ji}(s_1)|.
$$

By the Schwarz inequality

$$
\begin{aligned}
\left(\int_0^T |f_k(t-s) f_i(t-s_1)|\, dt\right)^2 &\le \int_0^T |f_k(t-s)|^2 dt \int_0^T |f_i(t-s_1)|^2 dt \\
&\le \int_{-\eta}^T |f_k(t)|^2 dt \int_{-\eta}^T |f_i(t)|^2 dt.
\end{aligned}
$$

Put

$$
a_{jk} = \int_0^{\eta} |w(s)||dr_{jk}(s)|.
$$

We obtain

$$
\int_0^T |(E_1 f)_j(t)|^2 dt \le \sum_{i=1}^{n} \sum_{k=1}^{n} a_{jk} a_{ik} \|f_k\|_{L^2(-\eta,T)} \|f_i\|_{L^2(-\eta,T)}
$$

$$= \left(\sum_{k=1}^{n} a_{jk} \|f_k\|_{L^2(-\eta,T)} \right)^2 .$$

Hence

$$\sum_{j=1}^{n} \int_0^T |(E_1 f)_j(t)|^2 dt \le \sum_{j=1}^{n} (\sum_{k=1}^{n} a_{jk} \|f_k\|_{L^2(-\eta,T)})^2 = \|(a_{jk})_{j,k=1}^n \nu_2\|_n$$
$$\le \|(a_{jk})_{j,k=1}^n\|_n \|\nu_2\|_n = Vd\ (R_0, w) \|\nu_2\|_n,$$

where

$$\nu_2 = \left(\|f_k\|_{L^2(-\eta,T)} \right)_{k=1}^n .$$

But $\|\nu_2\|_n = \|f\|_{L^2(-\eta,T)}$. So (8.6) is proved for a continuous function f but the set of continuous functions is dense in L^2. So the lemma is proved. □

Furthermore, introduce the norm

$$\|f\|_{\hat{C}([a,b],\mathbf{C}^n)} = [\sum_{k=1}^{n} \sup_{a \le t \le b} |f_k(t)|^2]^{1/2}$$

for a bounded function f. Recall that

$$\|f\|_{C([a,b],\mathbf{C}^n)} = \sup_{a \le t \le b} [\sum_{k=1}^{n} |f_k(t)|^2]^{1/2}.$$

Clearly, $\|f\|_{C([a,b],\mathbf{C}^n)} \le \|f\|_{\hat{C}([a,b],\mathbf{C}^n)}$ and

$$\|f\|^2_{\hat{C}([a,b],\mathbf{C}^n)} \le n \max_k \sup_{a \le t \le b} |f_k(t)|^2 \le n \sup_{a \le t \le b} \sum_{k=1}^{n} |f_k(t)|^2 = n \|f\|^2_{C([a,b],\mathbf{C}^n)}.$$

Lemma 1.8 *One has*

$$\|E_0 f\|_{\hat{C}([0,T],\mathbf{C}^n)} \le var(R_0) \|f\|_{\hat{C}([-\eta,T],\mathbf{C}^n)} \quad (T > 0).$$

Proof For each coordinate $(E_0 f)_j(t)$ of $E_0 f(t)$, $t \in [0, T]$, we have

$$|(E_0 f)_j(t)| = |\int_0^{\eta} \sum_{k=1}^{n} f_k(t-s) dr_{jk}(s)|$$

$$\leq \sum_{k=1}^{n} \int_{0}^{\eta} |dr_{jk}| \max_{0\leq s\leq \eta} |f_k(t-s)|$$

$$= \sum_{k=1}^{n} var(r_{jk}) \max_{0\leq s\leq \eta} |f_k(t-s)|.$$

So

$$\max_{0\leq t\leq T} |(E_0 f)_j(t)| \leq \sum_{k=1}^{n} var(r_{jk}) \max_{-\eta\leq t\leq T} |f_k(t)|.$$

Recall that ν_C is a vector whose coordinates are

$$\|f_k\|_{C(-\eta,T)},\ k=1,\ldots,n.$$

Then

$$\|E_0 f\|^2_{\hat{C}([0,T],\mathbf{C}^n)} = \sum_{j=1}^{n} \max_{0\leq t\leq T} |(E_0 f)_j(t)|^2 \leq \sum_{j=1}^{n} \left(\sum_{k=1}^{n} var(r_{jk}) \|f_k\|_{C(-\eta,T)} \right)^2$$

$$= \|Var(R_0)\nu_C\|_n^2 \leq \|Var(R_0)\|_n^2 \|\nu_C\|_n^2$$

$$= var^2(R_0) \|f\|^2_{\hat{C}([-\eta,T],\mathbf{C}^n)}.$$

As claimed. □

Now introduce the norm

$$\|f\|_{\hat{L}^\infty([a,b],\mathbf{C}^n)} = [\sum_{k=1}^{n} ess \sup_{a\leq t\leq b} |f_k(t)|^2]^{1/2}.$$

Recall that

$$\|f\|_{L^\infty([a,b],\mathbf{C}^n)} = ess \sup_{a\leq t\leq b} [\sum_{k=1}^{n} |f_k(t)|^2]^{1/2}.$$

Clearly, $\|f\|_{L^\infty([a,b],\mathbf{C}^n)} \leq \|f\|_{\hat{L}^\infty([a,b],\mathbf{C}^n)}$ and

$$\|f\|^2_{\hat{L}^\infty([a,b],\mathbf{C}^n)} \leq n \max_k ess \sup_{a\leq t\leq b} |f_k(t)|^2 \leq n\ ess \sup_{a\leq t\leq b} \sum_{k=1}^{n} |f_k(t)|^2$$

$$= n \|f\|^2_{L^\infty([a,b],\mathbf{C}^n)}.$$

From the previous lemma we easily obtain

Lemma 1.9 *One has*

$$\|E_0 f\|_{\hat{L}^\infty([0,T],\mathbf{C}^n)} \leq var(R_0)\|f\|_{\hat{L}^\infty([-\eta,T],\mathbf{C}^n)} \quad (T > 0).$$

1.9 Time-Variant Difference Operators

For a positive constant $T \leq \infty$ and a vector valued function $f \in C([-\eta, T], \mathbf{C}^n)$ or $f \in L^p([-\eta, T], \mathbf{C}^n)$ $(1 \leq p \leq \infty)$, put

$$Ef(t) = \int_0^\eta d_s R(t, s) f(t - s) \quad (0 \leq t \leq T), \tag{9.1}$$

where $R(t, s) = (r_{ij}(t, s))_{i,j=1}^n$ is a real $n \times n$-matrix-valued function defined on $[0, T] \times [0, \eta]$, which is continuous in t for each τ and

$$v_{jk} := \sup_{t \geq 0} var(r_{jk}(t, .)) < \infty \quad (j, k = 1, \ldots, n). \tag{9.2}$$

It is also assumed that $R(t, s)$ has a finite number of jumps in s.

Throughout the book operator E is defined similarly to operator E_0: in the case $f \in C([-\eta, T], \mathbf{C}^n)$ the integral for each t is understood as the Lebesgues-Stieljes integral. In the case of non-continuous $f \in L^p([-\eta, T], \mathbf{C}^n)$ we prove that E is bounded in the norm of L^p and extend to the whole space L^p.

Let $V(R)$ be the spectral norm of matrix $(v_{jk})_{j,k=1}^n$:

$$V(R) = \|(v_{jk})\|_n.$$

For $0 \leq t \leq T$ we have

$$\left| \int_0^\eta f_k(t - s) d_s r_{jk}(t, s) \right| \leq \max_{0 \leq s \leq \eta} |f_k(t - s)| \int_{-\eta}^T |d_s r_{jk}(t, s)|$$
$$\leq v_{jk} \|f_k\|_{C(-\eta,T)}.$$

Thus each coordinate $(Ef)_j(t)$ of $Ef(t)$ satisfies the inequality

$$|(Ef)_j(t)| = \left| \int_0^\eta \sum_{k=1}^n f_k(t - s) d_s r_{jk}(t, s) \right|$$
$$\leq \sum_{k=1}^n v_{jk} \|f_k\|_{C(-\eta,T)}.$$

Hence,

$$\sum_{j=1}^{n} |(Ef)_j(t)|^2 \le \sum_{j=1}^{n}\left(\sum_{k=1}^{n} v_{jk}\|f_k\|_{C(-\eta,T)}\right)^2$$
$$= \|(v_{jk})\nu_C\|_n^2 \le V(R)\|\nu_C\|_n^2,$$

where ν_C again is a vector whose coordinates are

$$\|f_k\|_{C(-\eta,T)}, k = 1, \dots, n.$$

As it was shown in the previous section,

$$\|\nu_C\|_n^2 \le n\|f\|^2_{C([-\eta,T],\mathbf{C}^n)}.$$

Thus,

$$\|Ef\|_{C([0,T],\mathbf{C}^n)} \le \sqrt{n}V(R)\|f\|_{C([-\eta,T],\mathbf{C}^n)} \quad (f \in C([-\eta, T], \mathbf{C}^n)). \tag{9.3}$$

Moreover,

$$\|Ef\|_{L^\infty([0,T],\mathbf{C}^n)} \le \sqrt{n}V(R)\|f\|_{L^\infty([-\eta,T],\mathbf{C}^n)} \quad (f \in L^\infty([-\eta, T], \mathbf{C}^n)) \tag{9.4}$$

for a continuous function f. But the set of continuous functions is dense in L^∞. So the previous inequality is valid on the whole space. Repeating the arguments of the proof of inequality (8.3) from the previous lemma, we obtain

$$\|Ef\|_{L^2([0,T],\mathbf{C}^n)} \le V(R)\|f\|_{L^2([-\eta,T],\mathbf{C}^n)} \quad (f \in L^2([-\eta, T], \mathbf{C}^n)). \tag{9.5}$$

Now let

$$f(t) = (f_k(t)) \in L^1([-\eta, T], \mathbf{C}^n).$$

Then

$$\int_0^T |(Ef)_j(t)|dt \le \sum_{k=1}^{n}\int_0^\eta\int_0^T |f_k(t-s)|dt|d_s r_{jk}(t,s)| \le \sum_{k=1}^{n} v_{jk}\int_{-\eta}^{T} |f_k(t)|dt.$$

So

$$\|Ef\|_{L^1([0,T],\mathbf{C}^n)} = \int_0^T \sqrt{\sum_{j=1}^{n}|(Ef)_j(t)|^2 dt} \le \int_0^T \sum_{j=1}^{n}|(Ef)_j(t)|dt$$

$$\le \sum_{j=1}^{n} \int_{-\eta}^{T} \sum_{k=1}^{n} v_{jk} |f_k(t)| dt$$

$$\le \sum_{j=1}^{n} \int_{-\eta}^{T} \sqrt{\sum_{k=1}^{n} v_{jk}^2 \sum_{k=1}^{n} |f_k(t)|^2} \; dt$$

Hence

$$\|Ef\|_{L^1([0,T],\mathbf{C}^n)} \le \varsigma(R)\|f\|_{L^1([-\eta,T],\mathbf{C}^n)} \quad (f \in L^1([-\eta, T], \mathbf{C}^n)), \tag{9.6}$$

where

$$\varsigma(R) = \sum_{j=1}^{n} \sqrt{\sum_{k=1}^{n} v_{jk}^2}.$$

We thus have proved the following result.

Lemma 1.10 *Suppose the entries $r_{jk}(t, s)$ of the matrix function $R(t, s)$ satisfy condition* (9.2). *Then the operator E defined by* (9.1) *is subject to the inequalities* (9.3)–(9.6).

Furthermore, according to inequality (8.5), we obtain

$$\sum_{j=1}^{n} \sqrt{\sum_{k=1}^{n} v_{jk}^2} \ge \sqrt{\sum_{j=1}^{n} \sum_{k=1}^{n} v_{jk}^2}.$$

That is,

$$\varsigma(R) \ge V(R).$$

Now the Riesz-Thorin theorem and previous lemma imply the following result.

Corollary 1.3 *Let condition* (9.2) *hold. Then for the operator E defined by* (9.1), *the following inequalities are valid:*

$$\|Ef\|_{L^p([0,T],\mathbf{C}^n)} \le \sqrt{n} V(R)\|f\|_{L^p([-\eta,T],\mathbf{C}^n)} \quad \text{if } p \ge 2$$

and

$$\|Ef\|_{L^p([0,T],\mathbf{C}^n)} \le \varsigma(R)\|f\|_{L^p([-\eta,T],\mathbf{C}^n)} \quad \text{if } 1 \le p < 2 \quad (f \in L^p([-\eta, T], \mathbf{C}^n)).$$

Recall that the norms $\|.\|_{\hat{C}([a,b],\mathbf{C}^n)}$ and $\|.\|_{\hat{L}^\infty([a,b],\mathbf{C}^n)}$ are introduced in the previous section.

Lemma 1.11 *One has*

$$\|Ef\|_{\hat{C}([0,T],\mathbf{C}^n)} \leq V(R)\|f\|_{\hat{C}([-\eta,T],\mathbf{C}^n)}$$

and

$$\|Ef\|_{\hat{L}^\infty([0,T],\mathbf{C}^n)} \leq V(R)\|f\|_{\hat{L}^\infty([-\eta,T],\mathbf{C}^n)}.$$

The proof of this lemma is similar to the proof of Lemma 1.8. It is left to the reader.

1.10 The Laplace Transform

Let $f(t)$ be a locally integrable function defined for all real numbers $t \geq 0$ and satisfying

$$|f(t)| \leq Me^{\delta t} \quad (t \geq 0)$$

with a positive constant M and a real constant δ. Then the Laplace transform of function f is the function $\hat{f}(s)$, defined by

$$\hat{f}(s) = \int_0^\infty e^{-st} f(t)dt.$$

The parameter s is a complex number s with $Re\, s > \delta$. For a continuous f, the inverse Laplace transform is given by the following complex integral,

$$f(t) = \frac{1}{2\pi i} \lim_{T\to\infty} \int_{c-Ti}^{c-Ti} e^{st}\hat{f}(s)ds = \frac{1}{2\pi i} \int_{c-i\infty}^{c-i\infty} e^{st}\hat{f}(s)ds,$$

which is known by various names (the Bromwich integral, the Fourier-Mellin integral, and Mellin's inverse formula). Here c is a real number so that the contour path of integration is in the region of convergence of $\hat{f}(s)$. If $f(t)$ is not continuous, then at each point t_0 of discontinuity t_0 one has

$$\frac{1}{2}(f(t_0+0) + f(t_0-0)) = \frac{1}{2\pi i} \int_{c-i\infty}^{c-i\infty} e^{st}\hat{f}(s)ds.$$

If the integrals

$$\int_0^\infty e^{\alpha t}|f(t)|dt \ (\alpha \in \mathbf{R})$$

and

$$\int_0^\infty e^{2\alpha t}|f(t)|^2 dt$$

converge, then the integral

$$\int_{-\infty}^\infty |\hat{f}(is-\alpha)|^2 ds$$

also converges, and

$$\int_0^\infty e^{2\alpha t}|f(t)|^2 dt = \frac{1}{2\pi}\int_{-\infty}^\infty |\hat{f}(is-\alpha)|^2 ds$$

cf. Doetsch (1961). Hence the Parseval equality

$$\int_0^\infty |f(t)|^2 dt = \frac{1}{2\pi}\int_{-\infty}^\infty |\hat{f}(is)|^2 ds$$

follows, provided the integrals

$$\int_0^\infty |f(t)| dt \text{ and } \int_0^\infty |f(t)|^2 dt$$

converge.

Similarly, if $h(t) = (h_k(t))$ is a vector valued function defined on the positive half-line, and

$$\|h(t)\|_n \le Me^{\alpha t} \quad (t \ge 0)$$

with the Euclidean norm, then the Laplace transform of $h(t)$ is the function $\hat{h}(s)$, defined by

$$\hat{h}(s) = \int_0^\infty e^{-st} h(t) dt.$$

The inverse transform is defined as in the scalar case.

If the integrals

$$\int_0^\infty e^{\alpha t} \|h(t)\|_n dt$$

and

$$\int_0^\infty e^{2\alpha t} \|h(t)\|_n^2 dt$$

converge, then the integral

$$\int_{-\infty}^\infty \|\hat{h}(is - \alpha)\|_n^2 ds$$

also converges, and

$$\int_0^\infty e^{2\alpha t} \|h(t)\|_n^2 dt = \frac{1}{2\pi} \int_{-\infty}^\infty \|\hat{h}(is - \alpha)\|_n^2 ds, \tag{10.1}$$

in particular the Parseval equality

$$\int_0^\infty \|h(t)\|_n^2 dt = \frac{1}{2\pi} \int_{-\infty}^\infty \|\hat{h}(is)\|_n^2 ds \tag{10.2}$$

is valid, provided

$$\int_0^\infty \|h(t)\|_n^2 dt < \infty \text{ and } \int_0^\infty \|h(t)\|_n dt < \infty.$$

Indeed, let $\hat{h}_k(s)$ be the Laplace transform of a coordinate $h_k(t)$ of $h(t)$. Then $\hat{h}(s) = (\hat{h}_k(s))$ and due to the Parseval equality for scalar functions

$$\begin{aligned}\int_0^\infty \|h(t)\|_n^2 dt &= \sum_{k=1}^n \int_0^\infty |h_k(t)|^2 dt = \sum_{k=1}^n \frac{1}{2\pi} \int_{-\infty}^\infty |\hat{h}_k(is)|^2 ds \\ &= \frac{1}{2\pi} \int_{-\infty}^\infty \|\hat{h}(is)\|_n^2 ds,\end{aligned}$$

and thus (10.2) is proved. Similarly, (10.1) can be proved.

Now let $F(t)$ be a locally integrable matrix valued function defined for all real numbers $t \geq 0$ and satisfying

$$\|F(t)\|_n \leq Me^{\delta t} \quad (t \geq 0)$$

Then the Laplace transform of function $F(t)$ is the function $\hat{F}(s)$, defined by

$$\hat{F}(s) = \int_0^\infty e^{-st} F(t)dt.$$

The inverse Laplace transform is given by the following complex integral,

$$\frac{1}{2\pi i} \int_{c-i\infty}^{c-i\infty} e^{st} \hat{F}(s)ds.$$

For any $v \in \mathbf{C}^n$ due to (10.2),

$$\int_0^\infty \|F(t)v\|_n^2 dt = \frac{1}{2\pi} \int_{-\infty}^\infty \|\hat{F}(is)v\|_n^2 ds. \tag{10.3}$$

Let $\{d_k\}$ be an arbitrary orthogonal normal basis in $\mathbf{C}^n$. Then

$$\sum_{k=1}^n \|F(t)d_k\|_n^2 = N_2^2(F(t))$$

and

$$\sum_{k=1}^n \|\hat{F}(s)d_k\|_n^2 = N_2^2(\hat{F}(s)).$$

Here and below $N_2(A)$ means the Hilbert-Schmidt (Frobenius) norm of a matrix A: $N_2^2(A) = \mathit{Trace}\ AA^*$. Taking into account (10.3) we get our next result.

Lemma 1.12 *One has*

$$\int_0^\infty N_2^2(F(t))dt = \frac{1}{2\pi} \int_{-\infty}^\infty N_2^2(\hat{F}(s))ds,$$

provided

$$\int_0^\infty N_2^2(F(t))dt < \infty \quad \text{and} \quad \int_0^\infty N_2(F(t))dt < \infty.$$

Similarly, the equality

$$\int_0^\infty e^{2\alpha t} N_2^2(F(t))dt = \frac{1}{2\pi} \int_{-\infty}^\infty N_2^2(\hat{F}(is - \alpha))ds$$

can be proved, provided

$$\int_0^\infty e^{2\alpha t} N_2^2(F(t))dt < \infty \quad \text{and} \quad \int_0^\infty e^{\alpha t} N_2(F(t))dt < \infty.$$

1.11 Comments

Sections 1.1–1.4: Here we present standard facts, mainly from the theory of Banach and ordered spaces, required in further consideration.

In Sects. 1.5 *and* 1.6 we followed Sects. I.9 and III.2 of the book (Daleckii and Krein 1971).

The Kantorovich generalized norm considered in *Sect.* 1.7 will be applied, in particular, in Chap. 7 for the stability analysis of nonlinear neutral type systems. As it was above mentioned, the generalized norm enables us to use information about equations more completely than a usual (number) norm.

Sections 1.8–1.9: From the theory of functions of a bounded variation (see for instance Hew itt and Stromberg (1969)), it is well known that for a fixed t, $R(t, \tau)$ can be represented as $R(t, \tau) = R_1(t, \tau) + R_2(t, \tau) + R_3(t, \tau)$, where $R_1(t, \tau)$ is a saltus function of bounded variation with at most countably many jumps on $[0, \eta]$, $R_2(t, \tau)$ is an absolutely continuous function and $R_3(t, \tau)$ is either zero or a singular function of bounded variation, i.e., $R_3(t, \tau)$ is non-constant, continuous and has the derivative in τ which is equal to zero almost everywhere on $[0, \eta]$. If $R_3(t, .)$ is not zero identically, then the expression

$$\int_0^\eta d_\tau R_3(t, \tau) f(t - \tau)$$

for a continuous function f, cannot be transformed to a Lebesgue integral or to a series. For a concrete example see Hewitt and Stromberg (1969, p. 457). But if R *has a finite number of jumps, then* $R_3(t, \tau)$ *is zero identically*. Because of this we assume that R and R_0 have finite numbers of jumps. The material of Sects. 1.8 and 1.9 is adopted from the book (Gil' 2013a, Sect. 1.12).

Chapter 2
Eigenvalues and Functions of Matrices

In this chapter we accumulate norm estimates for matrix-valued functions and bounds for the eigenvalues of matrices. In particular, we suggest estimates for the resolvents, powers of matrices as well as for matrix exponentials. We also present bounds for the spectral radius and investigate perturbations of eigenvalues. The material of this paper is systematically applied in the rest of the book chapters.

2.1 Some Definitions

Everywhere in this chapter, $\|x\|$ is the Euclidean norm of $x \in \mathbf{C}^n$: $\|x\| = \sqrt{(x,x)}$ with a scalar product $(.,.) = (.,.)_{C^n}$, I is the unit matrix.

For a linear operator A in $\mathbf{C}^n$ (matrix), $\lambda_k = \lambda_k(A)$ $(k = 1, \ldots, n)$ are the eigenvalues of A enumerated in an arbitrary order with their multiplicities, $\sigma(A)$ denotes the spectrum of A, A^* is the adjoint to A, and A^{-1} is the inverse to A; $R_\lambda(A) = (A - \lambda I)^{-1}$ $(\lambda \in \mathbf{C}, \lambda \notin \sigma(A))$ is the resolvent, $r_s(A)$ is the spectral radius, $\|A\| = \sup_{x \in \mathbf{C}^n} \|Ax\|/\|x\|$ is the (operator) spectral norm, $N_2(A)$ is the Hilbert-Schmidt (Frobenius) norm of A:

$$N_2(A) = \sqrt{Trace\ AA^*},$$

$A_I = (A - A^*)/2i$ is the imaginary component, $A_R = (A + A^*)/2$ is the real component,

$$\rho(A, \lambda) = \min_{k=1,\ldots,n} |\lambda - \lambda_k(A)|$$

is the distance between $\sigma(A)$ and a point $\lambda \in \mathbf{C}$; $\rho(A, C) = \min_k \inf_{s \in C} |s - \lambda_k(A)|$ is the distance between a contour C and $\sigma(A)$. $co(A)$ denotes the closed convex hull of $\sigma(A)$, $\alpha(A) = \max_k Re\ \lambda_k(A)$, $\beta(A) = \min_k Re\ \lambda_k(A)$; $r_l(A)$ is the lower spectral radius:

M.I. Gil', *Stability of Neutral Functional Differential Equations*,
Atlantis Studies in Differential Equations 3, DOI 10.2991/978-94-6239-091-1_2

$$r_l(A) = \min_{k=1,\dots,n} |\lambda_k(A)|.$$

In addition, $\mathbf{C}^{n\times n}$ is the set of all complex $n \times n$-matrices.

The following quantity plays an essential role in the sequel:

$$g(A) = (N_2^2(A) - \sum_{k=1}^{n} |\lambda_k(A)|^2)^{1/2}.$$

It sometimes called the departure from normality, cf. Stewart and Sun Ji-guang (1990). Since

$$\sum_{k=1}^{n} |\lambda_k(A)|^2 \geq |Trace\ A^2|,$$

one can write

$$g^2(A) \leq N_2^2(A) - |Trace\ A^2|. \tag{1.1}$$

In Sect. 2.2 of the book Gil' (2003) it is proved that

$$g^2(A) \leq 2N_2^2(A_I) = N_2^2(A - A^*)/2 \tag{1.2}$$

and

$$g(e^{i\tau}A + zI) = g(A) \tag{1.3}$$

for all $\tau \in \mathbf{R}$ and $z \in \mathbf{C}$. If A is a normal matrix: $AA^* = A^*A$, then $g(A) = 0$.

If A_1 and A_2 are commuting matrices, then

$$g(A_1 + A_2) \leq g(A_1) + g(A_2), \tag{1.4}$$

cf. Gil' (2003, Sect. 2.1). By the inequality between geometric and arithmetic mean values,

$$\left(\frac{1}{n}\sum_{k=1}^{n} |\lambda_k(A)|^2\right)^n \geq (\prod_{k=1}^{n} |\lambda_k(A)|)^2.$$

So

$$g^2(A) \leq N_2^2(A) - n(det\ A)^{2/n}. \tag{1.5}$$

By Schur's theorem (Marcus and Minc 1964, Sect. I.4.10.2), for any operator A in $\mathbf{C}^n$, there is an orthogonal normal basis (Schur's basis) $\{e_k\}_{k=1}^n$, in which A is represented by triangular matrix. That is,

$$Ae_k = \sum_{j=1}^{k} a_{jk} e_j \text{ with } a_{jk} = (Ae_k, e_j) \ \ (j = 1, \ldots, n),$$

and $a_{jj} = \lambda_j(A)$. So

$$A = D + V \ \ (\sigma(A) = \sigma(D)) \tag{1.6}$$

with a normal (diagonal) operator D defined by $De_j = \lambda_j(A)e_j$ $(j = 1, \ldots, n)$ and a nilpotent (strictly upper-triangular) operator V defined by

$$Ve_k = \sum_{j=1}^{k-1} a_{jk} e_j \ \ (k = 2, \ldots, n).$$

D and V will be called *the diagonal part* and *the nilpotent part* of A, respectively.

As it is shown in Gil' (2003, Lemma 2.3.2), the relation

$$g(A) = N_2(V) \tag{1.7}$$

is true. Hence it follows that

$$g(U^{-1}AU) = g(A) \text{ for any unitary matrix } U. \tag{1.8}$$

2.2 Representations of Matrix Functions

In this section we recall some classical representations of functions of matrices. For details see Horn and Johnson (1985, Chap. 6).

Let $A \in \mathbf{C}^{n\times n}$ and $M \supset \sigma(A)$ be an open simply-connected set whose boundary C consists of a finite number of rectifiable Jordan curves, oriented in the positive sense customary in the theory of complex variables. Suppose that $M \cup C$ is contained in the domain of analyticity of a scalar-valued function f. Then $f(A)$ can be defined by the generalized integral formula of Cauchy

$$f(A) = -\frac{1}{2\pi i}\int_C f(\lambda) R_\lambda(A) d\lambda. \tag{2.1}$$

If an analytic function $f(\lambda)$ is represented by the Taylor series

$$f(\lambda) = \sum_{k=0}^{\infty} c_k \lambda^k \ \ \left(|\lambda| < \frac{1}{\overline{\lim}_{k\to\infty} \sqrt[k]{|c_k|}}\right),$$

then one can define $f(A)$ as

$$f(A) = \sum_{k=0}^{\infty} c_k A^k$$

provided the spectral radius $r_s(A)$ of A satisfies the inequality

$$r_s(A)\overline{\lim}_{k\to\infty}\sqrt[k]{|c_k|} < 1.$$

In particular, for any matrix A, one has

$$e^A = \sum_{k=0}^{\infty} \frac{A^k}{k!}.$$

Consider the $n \times n$-Jordan block:

$$J_n(\lambda_0) = \begin{pmatrix} \lambda_0 & 1 & 0 & \dots & 0 \\ 0 & \lambda_0 & 1 & \dots & 0 \\ . & . & . & \dots & . \\ . & . & . & \dots & . \\ . & . & . & \dots & . \\ 0 & 0 & \dots & \lambda_0 & 1 \\ 0 & 0 & \dots & 0 & \lambda_0 \end{pmatrix},$$

then

$$f(J_n(\lambda_0)) = \begin{pmatrix} f(\lambda_0) & \frac{f'(\lambda_0)}{1!} & \dots & \frac{f^{(n-1)}(\lambda_0)}{(n-1)!} \\ 0 & f(\lambda_0) & \dots & \\ . & . & \dots & . \\ . & . & \dots & . \\ . & . & \dots & . \\ 0 & \dots & f(\lambda_0) & \frac{f'(\lambda_0)}{1!} \\ 0 & \dots & 0 & f(\lambda_0) \end{pmatrix}.$$

Thus, if A has the Jordan block-diagonal form

$$A = diag\ (J_{m_1}(\lambda_1), J_{m_2}(\lambda_2), \dots, J_{m_{n_0}}(\lambda_{n_0})),$$

where $\lambda_k, k = 1, \dots, n_0$ are the eigenvalues whose geometric multiplicities are m_k, then

$$f(A) = diag\ (f(J_{m_1}(\lambda_1)), f(J_{m_2}(\lambda_2)), \dots, f(J_{m_{n_0}}(\lambda_{n_0}))). \tag{2.2}$$

Note that in (2.2) we do not require that f is regular on a neighborhood of an open set containing $\sigma(A)$; one can take an arbitrary function which has at each λ_k derivatives up to $m_k - 1$-order.

In particular, if an $n \times n$-matrix A is diagonalizable, that is its eigenvalues have the geometric multiplicities $m_k \equiv 1$, then

$$f(A) = \sum_{k=1}^{n} f(\lambda_k) Q_k, \tag{2.3}$$

where Q_k are the eigenprojections. In the case (2.3) it is required only that f is defined on the spectrum.

Now let

$$\sigma(A) = \bigcup_{k=1}^{m} \sigma_k(A) \quad (m \le n)$$

and $\sigma_k(A) \subset M_k$ $(k = 1, \ldots, m)$, where M_k are open disjoint simply-connected sets: $M_k \cap M_j = \emptyset$ $(j \ne k)$. Let f_k be regular on M_k. Introduce on $M = \cup_{k=1}^{m} M_k$ the piece-wise analytic function by $f(z) = f_j(z)$ $(z \in M_j)$. Then

$$f(A) = -\frac{1}{2\pi i} \sum_{j=1}^{m} \int_{C_j} f(\lambda) R_\lambda(A) d\lambda, \tag{2.4}$$

where $C_j \subset M_j$ are closed smooth contours surrounding $\sigma(A_j)$ and the integration is performed in the positive direction. For more details about representation (2.4) see Yakubovich and Starzhinskii (1975, p. 49).

For instance, let M_1 and M_2 be two disjoint disks, and

$$f(z) = \begin{cases} \sin z & \text{if } z \in M_1, \\ \cos z & \text{if } z \in M_2 \end{cases}.$$

Then (2.4) holds with $m = 2$.

2.3 Norm Estimates for Resolvents

For a natural number $n \ge 2$, introduce the numbers

$$\gamma_{n,k} = \sqrt{\frac{(n-1)(n-2)\cdots(n-k)}{(n-1)^k k!}} \quad (k = 1, 2, \ldots, n-1), \ \gamma_{n,0} = 1.$$

Evidently, for all $n > 2$,

$$\gamma_{n,k}^2 \le \frac{1}{k!} \quad (k = 1, 2, \ldots, n-1). \tag{3.1}$$

Recall that $g(A)$ is defined in Sect. 2.1.

Theorem 2.1 *Let A be a linear operator in $\mathbf{C}^n$. Then its resolvent satisfies the inequality*

$$\|R_\lambda(A)\| \le \sum_{k=0}^{n-1} \frac{g^k(A)\gamma_{n,k}}{\rho^{k+1}(A,\lambda)} \quad (\lambda \notin \sigma(A)),$$

where $\rho(A,\lambda) = \min_{k=1,\ldots,n} |\lambda - \lambda_k(A)|$.

This theorem is proved in Gil' (2003, Theorem 2.1.1). Theorem 2.1 is sharp: if A is a normal matrix, then $g(A) = 0$ and Theorem 2.1 gives us the equality $\|R_\lambda(A)\| = 1/\rho(A,\lambda)$. Taking into account (3.1), we get

Corollary 2.1 *Let $A \in \mathbf{C}^{n\times n}$. Then*

$$\|R_\lambda(A)\| \le \sum_{k=0}^{n-1} \frac{g^k(A)}{\sqrt{k!}\rho^{k+1}(A,\lambda)} \textit{ for any regular } \lambda \textit{ of } A.$$

In particular, if A is invertible, then

$$\|A^{-1}\| \le \sum_{k=0}^{n-1} \frac{g^k(A)}{\sqrt{k!}r_l^{k+1}(A)},$$

where $r_l(A) = \rho(A,0) = \min_{k=1,\ldots,n} |\lambda_k(A)|$.

We will need also the following result.

Theorem 2.2 *Let $A \in \mathbf{C}^{n\times n}$. Then*

$$\|R_\lambda(A)\, det\,(\lambda I - A)\| \le \left[\frac{N_2^2(A) - 2Re\,(\bar\lambda\; Trace\,(A)) + n|\lambda|^2}{n-1}\right]^{(n-1)/2} \quad (\lambda \notin \sigma(A)).$$

In particular,

$$\|A^{-1}\| \le \frac{N_2^{n-1}(A)}{(n-1)^{(n-1)/2}|det\,(A)|}$$

for any invertible $A \in \mathbf{C}^{n\times n}$.

The proof of this theorem can be found in Gil' (2003, Sect. 2.11).

We also mention the following result.

Theorem 2.3 *Let $A \in \mathbf{C}^{n\times n}$. Then*

$$\|R_\lambda(A)\| \le \frac{1}{\rho(A,\lambda)}\left[1+\frac{1}{n-1}\left(1+\frac{g^2(A)}{\rho^2(A,\lambda)}\right)\right]^{(n-1)/2}$$

for any regular λ of A.

For the proof see Gil' (2003, Theorem 2.14.1).

2.4 Spectral Variations of Matrices

Let A and B be $n \times n$-matrices having eigenvalues

$$\lambda_1(A),\ldots,\lambda_n(A) \text{ and } \lambda_1(B),\ldots,\lambda_n(B),$$

respectively, and $q = \|A - B\|$.

The spectral variation of B with respect to A is

$$sv_A(B) := \max_i \min_j |\lambda_i(B) - \lambda_j(A)|,$$

cf. Stewart and Sun Ji-guang (1990).

The following simple lemma is proved in Gil' 2003, Sect. 4.1.

Lemma 2.1 *Assume that*

$$\|R_\lambda(A)\| \le \phi\left(\frac{1}{\rho(A,\lambda)}\right)$$

for all regular λ of A, where $\phi(x)$ is a monotonically increasing non-negative continuous function of a non-negative variable x, such that $\phi(0) = 0$ and $\phi(\infty) = \infty$. Then the inequality $sv_A(B) \le z(\phi, q)$ is true, where $z(\phi, q)$ is the unique positive root of the equation $q\phi(1/z) = 1$.

This Lemma and Corollary 2.1 yield our next result.

Theorem 2.4 *Let A and B be $n \times n$-matrices. Then $sv_A(B) \le z(q, A)$, where $z(q, A)$ is the unique nonnegative root of the algebraic equation*

$$y^n = q\sum_{j=0}^{n-1}\frac{y^{n-j-1}g^j(A)}{\sqrt{j!}}. \tag{4.1}$$

This theorem is sharp: if A is normal, then $g(A) = 0$ and $z(q, A) = q$. So the theorem gives us the inequality $sv_A(B) \leq q$. But $sv_A(B) = q$ in this case, cf. Stewart and Sun Ji-guang (1990).

Let us consider the algebraic equation

$$z^n = p(z) \quad (n > 1), \text{ where } p(z) = \sum_{j=0}^{n-1} c_j z^{n-j-1} \tag{4.2}$$

with non-negative coefficients c_j $(j = 0, \ldots, n-1)$.

Lemma 2.2 *The unique positive root z_0 of Eq. (4.2) is subject to the following estimates:*

$$z_0 \leq \sqrt[n]{p(1)} \text{ if } p(1) \leq 1, \tag{4.3}$$

and

$$1 \leq z_0 \leq p(1) \text{ if } p(1) \geq 1. \tag{4.4}$$

Proof Since all the coefficients of $p(z)$ are non-negative, it does not decrease as $z > 0$ increases. From this it follows that if $p(1) \leq 1$, then $z_0 \leq 1$. So $z_0^n \leq p(1)$, as claimed.

Now let $p(1) \geq 1$, then due to (4.2) $z_0 \geq 1$ because $p(z)$ does not decrease. It is clear that

$$p(z_0) \leq z_0^{n-1} p(1)$$

in this case. Substituting this inequality into (4.2), we get (4.4). □

Substituting $z = ax$ with a positive constant a into (4.2), we obtain

$$x^n = \sum_{j=0}^{n-1} \frac{c_j}{a^{j+1}} x^{n-j-1}. \tag{4.5}$$

Let

$$a = 2 \max_{j=0,\ldots,n-1} \sqrt[j+1]{c_j}.$$

Then

$$\sum_{j=0}^{n-1} \frac{c_j}{a^{j+1}} \leq \sum_{j=0}^{n-1} 2^{-j-1} = 1 - 2^{-n} < 1.$$

Let x_0 be the extreme right-hand root of Eq. (4.5), then by (4.3) we have $x_0 \leq 1$. Since $z_0 = ax_0$, we have derived the following result.

Corollary 2.2 *The unique nonnegative root* z_0 *of Eq.* (4.2) *satisfies the inequality*

$$z_0 \leq 2 \max_{j=0,\ldots,n-1} \sqrt[j+1]{c_j}.$$

Now put $y = xg(A)$ into (4.1). Then we obtain the equation

$$x^n = \frac{q}{g(A)} \sum_{j=0}^{n-1} \frac{x^{n-j-1}}{\sqrt{j!}}. \tag{4.6}$$

Since

$$\max_{j=0,\ldots,n-1} \sqrt[j+1]{\frac{q}{g(A)}} = \frac{q}{g(A)}$$

if $q \geq g(A)$, and

$$\max_{j=0,\ldots,n-1} \sqrt[j+1]{\frac{q}{g(A)}} = \sqrt[n]{\frac{q}{g(A)}}$$

if $q \leq g(A)$. Applying Corollary 2.2, we get the estimate $z(q, A) \leq \delta(q)$, where

$$\delta(q) := \begin{cases} 2q & \text{if } q \geq g(A) \\ 2g^{1-1/n}(A)q^{1/n} & \text{if } q \leq g(A) \end{cases}.$$

Now Theorem 2.4 ensures the following result.

Corollary 2.3 *One has*

$$sv_A(B) \leq \delta(q). \tag{4.7}$$

About other estimates for $sv_A(B)$ see Comments.

2.5 Norm Estimates for Matrix Functions

2.5.1 Estimates via the Resolvent

The following result directly follows from (2.1).

Lemma 2.3 *Let* $f(\lambda)$ *be a scalar-valued function which is regular on a neighborhood* M *of an open simply-connected set containing the spectrum of* $A \in \mathbf{C}^{n\times n}$, *and* $C \subset M$ *be a closed smooth contour surrounding* $\sigma(A)$. *Then*

$$\|f(A)\| \leq \frac{1}{2\pi} \int_C |f(z)| \|R_z(A)\| dz \leq m_C(A) l_C \sup_{z\in C} |f(z)|,$$

where

$$m_C(A) := \sup_{z \in C} \|R_z(A)\|, \ l_C := \frac{1}{2\pi} \int_C |dz|.$$

Now we can directly apply the estimates for the resolvent from Sect. 2.3. In particular, by Corollary 2.1 we have

$$\|R_z(A)\| \leq p(A, 1/\rho(A, z)), \tag{5.1}$$

where

$$p(A, x) = \sum_{k=0}^{n-1} \frac{x^{k+1} g^k(A)}{\sqrt{k!}} \quad (x > 0).$$

We thus get $m_C(A) \leq p(A, 1/\rho(A, C))$, where $\rho(A, C)$ is the distance between C and $\sigma(A)$, and therefore, we arrive at our next result.

Corollary 2.4 *Let $f(\lambda)$ be a scalar-valued function which is regular on a neighborhood M of an open simply-connected set containing the spectrum of $A \in \mathbf{C}^{n \times n}$, and $C \subset M$ be a closed smooth contour surrounding $\sigma(A)$. Then*

$$\|f(A)\| \leq b(A, C) \sup_{z \in C} |f(z)|,$$

where $b(A, C) = l_C p(A, 1/\rho(A, C)$. In particular,

$$\|A^m\| \leq b(A, C) \sup_{z \in C} |z|^m \quad (m = 1, 2, \ldots)$$

and

$$\|e^{At}\| \leq b(A, C) e^{\alpha_C t} \quad (t \geq 0),$$

where $\alpha_C = \sup_{z \in C} Re\, z$.

2.5.2 *Functions Regular on the Convex Hull of the Spectrum*

In this subsection, under additional conditions, we make the results of the previous subsection sharper although obtained here estimates are less convenient than the previous corollary.

Theorem 2.5 *Let A be an $n \times n$-matrix and f be a function holomorphic on a neighborhood of the convex hull $co(A)$ of $\sigma(A)$. Then*

$$\|f(A)\| \leq \sup_{\lambda\in\sigma(A)} |f(\lambda)| + \sum_{k=1}^{n-1} \sup_{\lambda\in co(A)} |f^{(k)}(\lambda)| \frac{\gamma_{n,k} g^k(A)}{k!}.$$

In particular

$$\|exp(At)\| \leq e^{\alpha(A)t} \sum_{k=0}^{n-1} g^k(A) t^k \frac{\gamma_{n,k}}{k!} \quad (t \geq 0)$$

where $\alpha(A) = \max_{k=1,\ldots,n} Re\ \lambda_k(A)$. *In addition,*

$$\|A^m\| \leq \sum_{k=0}^{n-1} \frac{\gamma_{n,k} m! g^k(A) r_s^{m-k}(A)}{(m-k)!k!} \quad (m = 1, 2, \ldots),$$

where $r_s(A)$ *is the spectral radius.*

Recall that $1/(m-k)! = 0$ if $m < k$.

This theorem is proved in the next subsection. It is a slight improvement of Theorem 2.7.1 from Gil' (2003).

Taking into account (3.1) we get our next result.

Corollary 2.5 *Under the hypothesis of Theorem 2.5 we have*

$$\|f(A)\| \leq \sup_{\lambda\in\sigma(A)} |f(\lambda)| + \sum_{k=1}^{n-1} \sup_{\lambda\in co(A)} |f^{(k)}(\lambda)| \frac{g^k(A)}{(k!)^{3/2}}.$$

In particular,

$$\|exp(At)\| \leq e^{\alpha(A)t} \sum_{k=0}^{n-1} \frac{g^k(A) t^k}{(k!)^{3/2}} \quad (t \geq 0)$$

and

$$\|A^m\| \leq \sum_{k=0}^{n-1} \frac{m! g^k(A) r_s^{m-k}(A)}{(m-k)!(k!)^{3/2}} \quad (m = 1, 2, \ldots).$$

Theorem 2.5 is sharp: if A is normal, then $g(A) = 0$ and

$$\|f(A)\| = \sup_{\lambda\in\sigma(A)} |f(\lambda)|.$$

2.5.3 Proof of Theorem 2.5

Let $|V|_e$ be the operator whose entries in the orthonormal basis of the triangular representation (the Schur basis) $\{e_k\}$ are the absolute values of the entries of the nilpotent part V of A with respect to this basis (see Sect. 2.1). That is,

$$|V|_e = \sum_{k=1}^{n} \sum_{j=1}^{k-1} |a_{jk}|(., e_k)e_j,$$

where $a_{jk} = (Ae_k, e_j)$. Put

$$I_{j_1 \dots j_{k+1}} = \frac{(-1)^{k+1}}{2\pi i} \int_C \frac{f(\lambda)d\lambda}{(\lambda_{j_1} - \lambda) \cdots (\lambda_{j_{k+1}} - \lambda)}.$$

We need the following result.

Lemma 2.4 *Let A be an $n \times n$-matrix and f be a holomorphic function on a Jordan domain (that is on a closed simply connected set, whose boundary is a Jordan contour), containing $\sigma(A)$. Let D be the diagonal part of A. Then*

$$\| f(A) - f(D)\| \leq \sum_{k=1}^{n-1} J_k \| \, |V|_e^k \|,$$

where

$$J_k = \max\{|I_{j_1 \dots j_{k+1}}| : 1 \leq j_1 < \cdots < j_{k+1} \leq n\}.$$

Proof From (2.1) and (1.6) we deduce that

$$f(A) - f(D) = -\frac{1}{2\pi i} \int_C f(\lambda)(R_\lambda(A) - R_\lambda(D))d\lambda = \sum_{k=1}^{n-1} B_k, \tag{5.2}$$

where

$$B_k = (-1)^{k+1} \frac{1}{2\pi i} \int_C f(\lambda)(R_\lambda(D)V)^k R_\lambda(D)d\lambda.$$

Since D is a diagonal matrix with respect to the Schur basis $\{e_k\}$ and its diagonal entries are the eigenvalues of A, then

$$R_\lambda(D) = \sum_{j=1}^{n} \frac{\Delta P_j}{\lambda_j(A) - \lambda},$$

where $\Delta P_k = (., e_k)e_k$. In addition, $\Delta P_j V \Delta P_k = 0$ for $j \geq k$. Consequently,

$$B_k = \sum_{j_1=1}^{j_2-1} \Delta P_{j_1} V \sum_{j_2=1}^{j_3-1} \Delta P_{j_2} V \cdots \sum_{j_k=1}^{j_{k+1}-1} V \sum_{j_{k+1}=1}^{n} \Delta P_{j_{k+1}} I_{j_1 j_2 \ldots j_{k+1}}.$$

Lemma 2.8.1 from Gil' (2003) gives us the estimate

$$\|B_k\| \leq J_k \| \sum_{j_1=1}^{j_2-1} \Delta P_{j_1} |V|_e \sum_{j_2=1}^{j_3-1} \Delta P_{j_2} |V|_e \cdots \sum_{j_k=1}^{j_{k+1}-1} |V|_e \sum_{j_{k+1}=1}^{n} \Delta P_{j_{k+1}} \|$$
$$= J_k \| P_{n-k} |V|_e P_{n-k+1} |V|_e P_{n-k+2} \cdots P_{n-1} |V|_e \|.$$

But

$$P_{n-k}|V|_e P_{n-k+1}|V|_e P_{n-k+2} \cdots P_{n-1}|V|_e = |V|_e P_{n-k+1}|V|_e P_{n-k+2} \cdots P_{n-1}|V|_e$$
$$= |V|_e^2 \cdots P_{n-1}|V|_e = |V|_e^k.$$

Thus

$$\|B_k\| \leq J_k \| \, |V|_e^k \, \|.$$

This inequality and (5.2) imply the required inequality. □

Thanks to Theorem 2.5.1 from the book (Gil' 2003), for any $n \times n$ nilpotent matrix V_0,

$$\|V_0^k\| \leq \gamma_{n,k} N_2^k(V_0) \ \ (k = 1, 2, \ldots, n-1). \tag{5.3}$$

Since $N_2(|V|_e) = N_2(V) = g(A)$, by (5.3) and the previous lemma we get the following result.

Lemma 2.5 *Under the hypothesis of Lemma 2.4 we have*

$$\|f(A) - f(D)\| \leq \sum_{k=1}^{n-1} J_k \gamma_{n,k} g^k(A).$$

Let f be holomorphic on a neighborhood of $co(A)$. Thanks to Lemma 1.5.1 from Gil' (2003),

$$J_k \leq \frac{1}{k!} \sup_{\lambda \in co(A)} |f^{(k)}(\lambda)|.$$

Now the previous lemma implies

Corollary 2.6 *Under the hypothesis of Theorem 2.5 we have the inequality*

$$\|f(A)-f(D)\| \leq \sum_{k=1}^{n-1} \sup_{\lambda \in co(A)} |f^{(k)}(\lambda)| \gamma_{n,k} \frac{g^k(A)}{k!}.$$

The assertion of Theorem 2.5 directly follows from the previous corollary.

Let us point to an additional estimate, which may be more convenient than Corollary 2.6 in the case when is n rather small. Denote by $f[a_1, a_2, \ldots, a_{k+1}]$ the k-th divided difference of f at points $a_1, a_2, \ldots, a_{k+1}$. By the Hadamard representation (Gel'fond 1967, formula (54)), we have

$$I_{j_1 \ldots j_{k+1}} = f[\lambda_1, \ldots, \lambda_{j_{k+1}}],$$

provided all the eigenvalues $\lambda_j = \lambda_j(A)$ are distinct. Now Lemma 2.4 implies

Corollary 2.7 *Let all the eigenvalues of an $n \times n$-matrix A be algebraically simple, and f be a holomorphic function in a Jordan domain containing $\sigma(A)$. Then*

$$\|f(A)-f(D)\| \leq \sum_{k=1}^{n-1} f_k \gamma_{n,k} g^k(A) \leq \sum_{k=1}^{n-1} f_k \frac{g^k(A)}{\sqrt{k!}},$$

where

$$f_k = \max\{|f[\lambda_1(A), \ldots, \lambda_{j_{k+1}}(A)]| : 1 < j_1 < \cdots < j_{k+1} \leq n\}.$$

2.6 Absolute Values of Elements of Matrix Functions

2.6.1 Statement of the Result

In this section we suggest bounds for the entries of a matrix function. Below we show that these bounds can be applied to the stability analysis via the generalized (vector) norms.

Everywhere in the present section, $A = (a_{jk})_{j,k=1}^n$, $S = diag\ [a_{11}, \ldots, a_{nn}]$ and the off diagonal of A is $W = A - S$. That is, the entries v_{jk} of W are $v_{jk} = a_{jk}$ $(j \neq k)$ and $v_{jj} = 0$ $(j, k = 1, 2, \ldots)$. Denote by $co(S)$ the closed convex hull of the diagonal entries $a_{11}, \ldots, a_{nn}$. We put $|A| = (|a_{jk}|)_{j,l=1}^n$, i.e. $|A|$ is the matrix whose entries are the absolute values of the entries A in the standard basis. We also write $T \geq 0$ if all the entries of a matrix T are nonnegative. If T and B are two matrices, then we write $T \geq B$ if $T - B \geq 0$.

Thanks to the Gerschgorin bound for the eigenvalues (see Sect. 2.11) we have $r_s(|W|) \leq \tau_W$, where

$$\tau_W := \max_j \sum_{k=1}^{n} |a_{jk}|.$$

Theorem 2.6 *Let $f(\lambda)$ be holomorphic on a neighborhood of a Jordan set, whose boundary C has the property*

$$|z - a_{jj}| > \sum_{k=1}^{n} |a_{jk}| \tag{6.1}$$

for all $z \in C$ and $j = 1, \ldots, n$. Then, with the notation

$$\xi_k(A) := \sup_{z \in co\,(S)} \frac{|f^{(k)}(z)|}{k!} \quad (k = 1, 2, \ldots),$$

the inequality

$$|f(A) - f(S)| \le \sum_{k=1}^{\infty} \xi_k(A)|W|^k$$

is valid, provided

$$r_s(|W|)\overline{\lim}_{k\to\infty} \sqrt[k]{\xi_k(A)} < 1.$$

This theorem is proved in the next subsection.

Corollary 2.8 *Under the hypothesis of the previous theorem with the notation*

$$\xi_0(A) := \max_k |f(a_{kk})|,$$

we have the inequality

$$|f(A)| \le \xi_0(A)I + \sum_{k=1}^{\infty} \xi_k(A)|W|^k = \sum_{k=1}^{\infty} \xi_k(A)|W|^k.$$

Here $|W|^0 = I$. In particular,

$$|A^m| \le |A|^m \le \sum_{k=1}^{m} \frac{m!}{(m-k)!k!} \max_j |a_{jj}|^{m-k}|W|^k = (\max_j |a_{jj}| + |W|)^m$$

and

$$|e^{At}| \le e^{\alpha(S)t} \sum_{k=1}^{\infty} \frac{t^k}{k!}|W|^k = e^{(\alpha(S)I+|W|)t} \quad (t > 0)$$

where $\alpha(S) = \max_k Re\ a_{kk}$.

Let $\|A\|_l$ denote a lattice norm of A. That is, $\|A\|_l \leq \||A|\|_l$, and $\|A\|_l \leq \|\tilde{A}\|_l$ whenever $0 \leq A \leq \tilde{A}$. Now the previous theorem implies the inequality

$$\|f(A) - f(S)\|_l \leq \sum_{k=1}^{\infty} \xi_k(A)\||W|^k\|_l$$

and therefore,

$$\|f(A)\|_l \leq \sum_{k=1}^{\infty} \xi_k(A)\||W|^k\|_l.$$

Additional estimates for the entries of matrix functions can be found in the paper (Gil' 2013e) and references given therein.

2.6.2 *Proof of Theorem 2.6*

By the equality $A = S + W$ we get

$$\begin{aligned} R_\lambda(A) &= (S + W - \lambda I)^{-1} = (I + R_\lambda(S)W)^{-1} R_\lambda(S) \\ &= \sum_{k=0}^{\infty} (R_\lambda(S)W)^k (-1)^k R_\lambda(S), \end{aligned}$$

provided the spectral radius $r_0(\lambda)$ of $R_\lambda(S)W$ is less than one. The entries of this matrix are

$$\frac{a_{jk}}{a_{jj} - \lambda} \quad (\lambda \neq a_{jj}, \quad j \neq k)$$

and the diagonal entries are zero. Thanks to (6.1) and the Gerschgorin bound for the eigenvalues from Sect. 2.11 of the present book, we have

$$r_0(\lambda) \leq \max_j \sum_{k=1}^{n} \frac{|a_{jk}|}{|a_{jj} - \lambda|} < 1 \quad (\lambda \in C)$$

and therefore, the series

$$R_\lambda(A) - R_\lambda(S) = \sum_{k=1}^{\infty} (R_\lambda(S)W)^k (-1)^k R_\lambda(S)$$

converges. Thus

$$f(A) - f(S) = -\frac{1}{2\pi i} \int\limits_C f(\lambda)(R_\lambda(A) - R_\lambda(S))d\lambda = \sum_{k=1}^{\infty} M_k, \tag{6.2}$$

where

$$M_k = (-1)^{k+1} \frac{1}{2\pi i} \int\limits_C f(\lambda)(R_\lambda(S)W)^k R_\lambda(S)d\lambda.$$

Since S is a diagonal matrix with respect to the standard basis $\{d_k\}$, we can write out

$$R_\lambda(S) = \sum_{j=1}^{n} \frac{\hat{Q}_j}{b_j - \lambda} \quad (b_j = a_{jj}),$$

where $\hat{Q}_k = (., d_k)d_k$. We thus have

$$M_k = \sum_{j_1=1}^{n} \hat{Q}_{j_1} W \sum_{j_2=1}^{n} \hat{Q}_{j_2} W \cdots W \sum_{j_{k+1}=1}^{n} \hat{Q}_{j_{k+1}} J_{j_1 j_2 \cdots j_{k+1}}. \tag{6.3}$$

Here

$$J_{j_1 \cdots j_{k+1}} = \frac{(-1)^{k+1}}{2\pi i} \int\limits_C \frac{f(\lambda)d\lambda}{(b_{j_1} - \lambda) \cdots (b_{j_{k+1}} - \lambda)}.$$

Lemma 1.5.1 from Gil' (2003) gives us the inequalities

$$|J_{j_1 \cdots j_{k+1}}| \leq \xi_k(A) \quad (j_1, j_2, \ldots, j_{k+1} = 1, \ldots, n).$$

Hence, by (6.3)

$$|M_k| \leq \xi_k(A) \sum_{j_1=1}^{n} \hat{Q}_{j_1} |W| \sum_{j_2=1}^{n} \hat{Q}_{j_2} |W| \cdots |W| \sum_{j_{k+1}=1}^{n} \hat{Q}_{j_{k+1}}.$$

But

$$\sum_{j_1=1}^{n} \hat{Q}_{j_1} |W| \sum_{j_2=1}^{n} \hat{Q}_{j_2} |W| \cdots |W| \sum_{j_{k+1}=1}^{n} \hat{Q}_{j_{k+1}} = |W|^k.$$

Thus $|M_k| \leq \xi_k(A)|W|^k$. Now (6.2) implies the required result. □

2.7 Diagonalizable Matrices

Everywhere in this section it is assumed that the eigenvalues $\lambda_k = \lambda_k(A)$ $(k = 1, \ldots, n)$ of A, taken with their algebraic multiplicities, are geometrically simple. That is, the geometric multiplicity of each eigenvalue is equal to one. As it is well-known, in this case A is diagonalizable: there are biorthogonal sequences $\{u_k\}$ and $\{v_k\}$: $(v_j, u_k) = 0$ $(j \neq k)$, $(v_j, u_j) = 1$ $(j, k = 1, \ldots, n)$, such that

$$A = \sum_{k=1}^{n} \lambda_k Q_k, \tag{7.1}$$

where $Q_k = (., u_k)v_k$ $(k = 1, \ldots, n)$ are one dimensional eigenprojections. Besides, there is an invertible operator T and a normal operator S, such that

$$TA = ST. \tag{7.2}$$

The constant (the condition number)

$$\kappa(A, T) = \kappa(A) := \|T\|\|T^{-1}\|$$

is very important for various applications, cf. Stewart and Sun Ji-guang (1990). That constant is mainly numerically calculated.

Making use equality (7.2) it is not hard to prove the following lemma.

Lemma 2.6 *Let A be a diagonalizable $n \times n$-matrix and $f(z)$ be a scalar function defined on the spectrum of A. Then $\|f(A)\| \leq \kappa(A) \max_k |f(\lambda_k)|$.*

In particular, we have

$$\|R_z(A)\| \leq \frac{\kappa(A)}{\rho(A, \lambda)},$$

$$\|e^{At}\| \leq \kappa(A)e^{\alpha(A)t} \quad (t \geq 0) \tag{7.3}$$

and

$$\|A^m\| \leq \kappa(A)r_s^m(A) \quad (m = 1, 2, \ldots).$$

Let A and $\tilde{A}$ be complex $n \times n$-matrices whose eigenvalues λ_k and $\tilde{\lambda}_k$, respectively, are taken with their algebraic multiplicities. Recall that

$$sv_A(\tilde{A}) := \max_k \min_j |\tilde{\lambda}_k - \lambda_j|.$$

Corollary 2.9 *Let A be diagonalizable. Then $sv_A(\tilde{A}) \leq \kappa(A)\|A - \tilde{A}\|$.*

Indeed, the operator $S = TAT^{-1}$ is normal. Put $B = T\tilde{A}T^{-1}$. Thanks to the well-known Corollary 3.4 from Stewart and Sun Ji-guang (1990), one has $sv_S(B) \le \|S - B\|$. Hence we get the required result.

Furthermore let $\tilde{D}$, V_+ and V_- be the diagonal, upper nilpotent part and lower nilpotent part of matrix $A = (a_{kk})$, respectively. Using the preceding corollary with $A_+ = \tilde{D} + V_+$, we arrive at the relations

$$\sigma(A_+) = \sigma(\tilde{D}), \text{ and } \|A - A_+\| = \|V_-\|.$$

Due to the previous corollary we get

Corollary 2.10 *Let $A = (a_{jk})_{j,k=1}^n$ be an $n \times n$-matrix, whose diagonal has the property*

$$a_{jj} \ne a_{kk} \ (j \ne k;\ k = 1, \dots, n).$$

Then for any eigenvalue μ of A, there is a $k = 1, \dots, n$, such that

$$|\mu - a_{kk}| \le \kappa(A_+)\|V_-\|,$$

and therefore the (upper) spectral radius satisfies the inequality

$$r_s(A) \le \max_{k=1,\dots,n} |a_{kk}| + \kappa(A_+)\|V_-\|,$$

and the lower spectral radius satisfies the inequality

$$r_s(A) \ge \min_{k=1,\dots,n} |a_{kk}| - \kappa(A_+)\|V_-\|,$$

provided $|a_{kk}| > \kappa(A_+)\|V_-\|$ $(k = 1, \dots, n)$.

Clearly, one can exchange the places of V_+ and V_-.

Let us point an estimate for the condition number $\kappa(A)$ in the case

$$\lambda_j(A) \ne \lambda_m(A) \text{ whenever } j \ne m. \tag{7.4}$$

To this end put

$$\delta := \min_{j \ne m} |\lambda_j(A) - \lambda_m(A)|, \ \ \tau(A) := \sum_{k=1}^{n-2} \frac{g^{k+1}(A)}{\sqrt{k!}\delta^{k+1}}$$

and

$$\gamma(A) := \left(1 + \frac{\tau(A)}{\sqrt{n-1}}\right)^{2(n-1)}.$$

Theorem 2.7 *Let condition* (7.4) *be fulfilled. Then there is an invertible matrix* T, *such that* (7.2) *holds with* $\kappa(A) \leq \gamma(A)$.

The proof of this theorem is presented in Gil' (2014). It is sharp: if A is normal, then $g(A) = 0$ and $\gamma(A) = 1$. Thus we obtain the equality $\kappa(A) = 1$.

2.8 Perturbations of Matrix Exponentials

The matrix exponential plays an essential role in the stability analysis of NDEs whose regular parts are close to ordinary differential equations.

As it was shown in Sect. 2.5,

$$\|exp(At)\| \leq e^{\alpha(A)t} \sum_{k=0}^{n-1} g^k(A) t^k \frac{\gamma_{n,k}}{k!} \quad (t \geq 0)$$

where $\alpha(A) = \max_{k=1,\dots,n} Re\ \lambda_k(A)$. Moreover, by (3.1),

$$\|exp(At)\| \leq e^{\alpha(A)t} \sum_{k=0}^{n-1} \frac{g^k(A) t^k}{(k!)^{3/2}} \quad (t \geq 0). \tag{8.1}$$

Taking into account that the operator $exp(-At)$ is the inverse one to $exp(At)$ it is not hard to show that

$$\|exp(At)h\| \geq \frac{e^{\beta(A)t}\|h\|}{\sum_{k=0}^{n-1} g^k(A) t^k (k!)^{-1} \gamma_{n,k}} \quad (t \geq 0, h \in \mathbf{C}^n),$$

where $\beta(A) = \min_{k=1,\dots,n} Re\ \lambda_k(A)$. Therefore by (3.1),

$$\|exp(At)h\| \geq \frac{e^{\beta(A)t}\|h\|}{\sum_{k=0}^{n-1} g^k(A)(k!)^{-3/2}\, t^k} \quad (t \geq 0). \tag{8.2}$$

Moreover, if A is a diagonalizable $n \times n$-matrix, then due to (7.3) we conclude that

$$\|e^{-At}\| \leq \kappa(A) e^{-\beta(A)t} \quad (t \geq 0).$$

Hence,

$$\|e^{At}h\| \geq \frac{\|h\| e^{\beta(A)t}}{\kappa(A)} \quad (t \geq 0, h \in \mathbf{C}^n).$$

Let $A, \tilde{A} \in \mathbf{C}^{n\times n}$ and $E = \tilde{A} - A$. We will say that A is stable (Hurwitzian), if $\alpha(A) < 0$. Assume that A is stable and put

$$u(A) = \int_0^\infty \|e^{At}\| dt.$$

To investigate perturbations of matrix exponentials one can use the identity

$$e^{\tilde{A}t} - e^{At} = \int_0^t e^{\tilde{A}(t-s)} E e^{As} ds. \tag{8.3}$$

Hence,

$$\int_0^T \|e^{\tilde{A}t} - e^{At}\| dt \leq \|E\| \int_0^T \int_0^t \|e^{\tilde{A}s}\| \|e^{A(t-s)}\| ds dt.$$

Consequently,

$$\int_0^T \|e^{\tilde{A}t}\| dt \leq u(A) \|E\| \int_0^T \|e^{\tilde{A}s}\| \int_s^T \|e^{A(t-s)}\| dt ds.$$

But

$$\int_s^T \|e^{A(t-s)}\| dt = \int_0^{T-s} \|e^{At_1}\| dt_1 \leq u(A).$$

So

$$\int_0^T \|e^{\tilde{A}t}\| dt \leq u(A) + u(A) \|E\| \int_0^T \|e^{\tilde{A}s}\| ds.$$

We thus arrive at

Theorem 2.8 *Let A be stable, and*

$$\|E\| u(A) < 1.$$

Then $\tilde{A}$ is also stable. Moreover,

$$u(\tilde{A}) \leq \frac{u(A)}{1 - u(A)\|E\|}.$$

Denote

$$v_A = \int_0^\infty t\|e^{At}\|dt.$$

Here we also investigate the perturbations in the case when

$$\|\tilde{A}E - EA\|v_A < \|E\|u(A).$$

Theorem 2.9 *Let A be stable, and*

$$\|\tilde{A}E - EA\|v_A < 1. \tag{8.4}$$

Then $\tilde{A}$ is also stable. Moreover,

$$u(\tilde{A}) \le \frac{u(A) + v_A\|E\|}{1 - v_A\|\tilde{A}E - EA\|} \tag{8.5}$$

and

$$\int_0^\infty \|e^{\tilde{A}t} - e^{At}\|dt \le \|E\|v_A + \frac{\|\tilde{A}E - EA\|v_A(u(A) + v_A\|E\|)}{1 - v_A\|\tilde{A}E - EA\|}. \tag{8.6}$$

This theorem is proved in this section below.

For example, if A and $\tilde{A}$ commute, then A and E commute, and $\tilde{A}E - EA = E^2$. If, in addition $E^2 = 0$, then $\tilde{A}E - EA = 0$. In this case Theorem 2.9 is sharper than Theorem 2.8.

Furthermore, by (8.1) we obtain $u(A) \le u_0(A)$ and $v_A \le \hat{v}_A$, where

$$u_0(A) := \sum_{k=0}^{n-1} \frac{g^k(A)}{|\alpha(A)|^{k+1}(k!)^{1/2}}$$

and

$$\hat{v}_A := \sum_{k=0}^{n-1} \frac{(k+1)g^k(A)}{|\alpha(A)|^{k+2}(k!)^{1/2}}.$$

Thus, Theorem 2.9 implies

Corollary 2.11 *Let A be stable and $\|\tilde{A}E - EA\|\hat{v}_A < 1$. Then $\tilde{A}$ is also stable. Moreover,*

$$u(\tilde{A}) \le \frac{u_0(A) + \hat{v}_A\|E\|}{1 - \hat{v}_A\|\tilde{A}E - EA\|}$$

and

$$\int_0^\infty \|e^{\tilde{A}t} - e^{At}\|dt \le \|E\|\hat{v}_A + \frac{\|\tilde{A}E - EA\|\hat{v}_A(u_0(A) + \hat{v}_A\|E\|)}{1 - \hat{v}_A\|\tilde{A}E - EA\|}.$$

Proof of Theorem 2.9: We use the following result, let $f(t)$, $c(t)$ and $h(t)$ be matrix functions defined on $[0, b]$ $(0 < b < \infty)$. Besides, f and h are differentiable and c is integrable. Then

$$\int_0^b f(t)c(t)h(t)dt = f(b)j(b)h(b) - \int_0^b (f'(t)j(t)h(t) + f(t)j(t)h'(t))dt \quad (8.7)$$

with

$$j(t) = \int_0^t c(s)ds.$$

Indeed, clearly,

$$\frac{\mathrm{d}}{\mathrm{d}t} f(t)j(t)h(t) = f'(t)j(t)h(t) + f(t)c(t)h(t) + f(t)j(t)h'(t).$$

Integrating this equality and taking into account that $j(0) = 0$, we arrive at (8.7).

By (8.7)

$$\begin{aligned} e^{\tilde{A}t} - e^{At} &= \int_0^t e^{A(t-s)} E e^{As} \mathrm{d}s \\ &= Ete^{At} + \int_0^t e^{\tilde{A}(t-s)}[\tilde{A}E - EA]se^{As}\mathrm{d}s. \end{aligned}$$

Hence,

$$\int_0^\infty \|e^{\tilde{A}t} - e^{At}\|\mathrm{d}t \le \int_0^\infty \|Ete^{At}\|\mathrm{d}t + \int_0^\infty \int_0^t \|e^{\tilde{A}(t-s)}\| \|\tilde{A}E - EA\| \|se^{As}\|\mathrm{d}s\ \mathrm{d}t.$$

But

$$\int_0^\infty \int_0^t \|e^{\tilde{A}(t-s)}\| s \|e^{As}\| ds\, dt = \int_0^\infty \int_s^\infty \|e^{\tilde{A}(t-s)}\| s \|e^{As}\| dt\, ds$$
$$= \int_0^\infty s\|e^{As}\| ds \int_0^\infty \|e^{\tilde{A}t}\| dt = \nu_A u(\tilde{A}).$$

Thus

$$\int_0^\infty \|e^{\tilde{A}t} - e^{At}\| dt \le \|E\|\nu_A + \|\tilde{A}E - EA\|\nu_A u(\tilde{A}). \tag{8.8}$$

Hence,

$$u(\tilde{A}) \le u(A) + \|E\|\nu_A + \|\tilde{A}E - EA\|\nu_A u(\tilde{A}).$$

So according to (8.4), we get (8.5). Furthermore, due to (8.8) and (8.5) we get (8.6). As claimed. □

2.9 Functions of Matrices with Nonnegative Off-Diagonals

In this section we establish two-sided inequalities for functions of matrices with nonnegative off-diagonals. They improve the results of Sect. 2.6 for these matrices and can be applied to the stability analysis of NDEs whose matrix coefficients have nonnegative off-diagonals.

In this section it is assumed that $A = (a_{ij})_{j,k=1}^n$ is a real matrix with

$$a_{ij} \ge 0 \text{ for } i \ne j. \tag{9.1}$$

Put

$$a = \min_{j=1,\dots,n} a_{jj} \text{ and } b = \max_{j=1,\dots,n} a_{jj}.$$

For a scalar function $f(\lambda)$ denote

$$\alpha_k(f, A) := \inf_{a \le x \le b} \frac{f^{(k)}(x)}{k!}$$

and

$$\beta_k(f, A) := \sup_{a \le x \le b} \frac{f^{(k)}(x)}{k!} \quad (k = 0, 1, 2, \ldots),$$

assuming that the derivatives exist.

Let $W = A - diag\ (a_{jj})$ be the off diagonal part of A.

Theorem 2.10 *Let condition (9.1) hold and $f(\lambda)$ be holomorphic on a neighborhood of a Jordan set, whose boundary C has the property*

$$|z - a_{jj}| > \sum_{k=1}^{n} a_{jk}$$

for all $z \in C$ and $j = 1, \ldots, n$. In addition, let f be real on $[a, b]$. Then the following inequalities are valid:

$$f(A) \ge \sum_{k=1}^{\infty} \alpha_k(f, A) W^k, \tag{9.2}$$

provided

$$r_s(W)\overline{\lim}_{k\to\infty} \sqrt[k]{|\alpha_k(f, A)|} < 1,$$

and

$$f(A) \le \sum_{k=1}^{\infty} \beta_k(f, A) W^k, \tag{9.3}$$

provided,

$$r_s(W)\overline{\lim}_{k\to\infty} \sqrt[k]{|\beta_k(f, A)|} < 1.$$

In particular, if $\alpha_k(f, A) \ge 0 \quad (k = 0, 1, \ldots)$, then matrix $f(A)$ has nonnegative entries.

Proof By (6.2) and (6.3),

$$f(A) = f(S) + \sum_{k=1}^{\infty} M_k,$$

where

$$M_k = \sum_{j_1=1}^{n} \hat{Q}_{j_1} W \sum_{j_2=1}^{n} \hat{Q}_{j_2} W \ldots W \sum_{j_{k+1}=1}^{n} \hat{Q}_{j_{k+1}} J_{j_1 j_2 \ldots j_{k+1}}.$$

Here

$$J_{j_1 \dots j_{k+1}} = \frac{(-1)^{k+1}}{2\pi i} \int_C \frac{f(\lambda) d\lambda}{(b_{j_1} - \lambda) \dots (b_{j_{k+1}} - \lambda)} \quad (b_j = a_{jj}).$$

Since S is real, Lemma 1.5.2 from Gil' (2003) gives us the inequalities

$$\alpha_k(f, A) \le J_{j_1 \dots j_{k+1}} \le \beta_k(f, A).$$

Hence,

$$M_k \ge \alpha_k(f, A) \sum_{j_1=1}^{n} \hat{Q}_{j_1} W \sum_{j_2=1}^{n} \hat{Q}_{j_2} W \dots W \sum_{j_{k+1}=1}^{n} \hat{Q}_{j_{k+1}} = \alpha_k(f, A) W^k.$$

Similarly, $M_k \le \beta_k(f, A) W^k$. This implies the required result. □

In addition, the previous theorem implies

$$e^{\min_j a_{jj} t} \sum_{k=1}^{\infty} \frac{t^k}{k!} W^k \le e^{At} \le e^{\max_j a_{jj} t} \sum_{k=1}^{\infty} \frac{t^k}{k!} W^k \ (t > 0)$$

Thus we arrive at the following

Corollary 2.12 *Let condition* (9.1) *hold. Then the following inequalities are valid:*

$$\sum_{k=1}^{m} \frac{m!}{(m-k)!k!} \min_j a_{jj}^{m-k} W^k \le A^m \le \sum_{k=1}^{m} \frac{m!}{(m-k)!k!} \max_j a_{jj}^{m-k} W^k$$

and

$$e^{(\min_j a_{jj} + W)t} \le e^{At} \le e^{(\max_j a_{jj} + W)t} \ (t > 0).$$

If all the diagonal elements are nonnegative then we obtain the trivial inequality.

$$(\min_j a_{jj} + W)^m \le A^m \le (\max_j a_{jj} + W)^m$$

If $max_j a_{jj} \ge 0$, then $A^m \le (\max_j a_{jj} + W)^m$.

2.10 Perturbations of Determinants

In this section $|||A|||$ is an arbitrary matrix norm and $\|A\|$ again denotes the spectral norm (the operator norm with respect to the Euclidean vector norm) of an $n \times n$ matrix A.

The following inequality for the determinants of two $n \times n$-matrices A and B is well-known (Bhatia 2007, p. 107):

$$|det\ A - det\ B| \leq n M_2^{n-1} \|A - B\|, \tag{10.1}$$

where $M_2 := \max\{\|A\|, \|B\|\}$. The spectral norm is unitarily invariant, but often is not easy to compute that norm. Now we are going to derive the similar inequalities for non-spectral norms.

It is supposed that for a given matrix norm, there is a constant α_n independent of A, such that

$$|det\ A| \leq \alpha_n |||A|||^n. \tag{10.2}$$

The aim of this paper is to prove the following result.

Theorem 2.11 *Let A and B be $n \times n$-matrices and condition* (10.2) *hold. Then*

$$|det\ A - det\ B| \leq \gamma_n\ |||A - B|||\ (|||A - B||| + |||A + B|||)^{n-1}, \tag{10.3}$$

where

$$\gamma_n := \frac{\alpha_n n^n}{2^{n-1}(n-1)^{n-1}}.$$

The proof of this theorem is presented in this section below.

For the spectral norm inequality (10.3), can be worse than (10.1). Indeed, for the spectral norm we have $\alpha_n = 1$. If we take $A = aB$ with a positive constant $a < 1$, then $\|A - B\| + \|A + B\| = 2\|B\| = 2M_2$, but

$$\gamma_n 2^{n-1} = n(1 + \frac{1}{n-1})^{n-1} \geq n.$$

Furthermore, let

$$N_p(A) := (Trace(A^* A)^{p/2})^{1/p} \ \ (p \geq 1)$$

be the Schatten-von Neumann norm In particular, $N_2(A)$ is the Frobenius (Hilbert-Schmidt) norm. Due to the inequality between the arithmetic and geometric mean values,

$$|det\ A|^p = \prod_{k=1}^{n} |\lambda_k(A)|^p \leq \left(\frac{1}{n} \sum_{k=1}^{n} |\lambda_k(A)|^p\right)^n.$$

Thus,

$$|det\ A| \le \frac{1}{n^{n/p}} N_p^n(A).$$

So in this case $\alpha_n = \frac{1}{n^{n/p}}$ and $\gamma_n = \zeta_{n,p}$, where

$$\zeta_{n,p} := \frac{n^n}{2^{n-1} n^{n/p} (n-1)^{n-1}}.$$

Now (10.3) implies

Corollary 2.13 *One has*

$$|det\ A - det\ B| \le \zeta_{n,p} N_p(A-B)\,(N_p(A-B) + N_p(A+B))^{n-1}. \tag{10.4}$$

Proof of Theorem 2.11: Let X and Y be complex normed spaces with norms $\|.\|_X$ and $\|.\|_Y$, respectively, and F be a Y-valued function defined on X. Assume that $F(C + \lambda \tilde{C})$ $(\lambda \in \mathbf{C})$ is an entire function for all $C, \tilde{C} \in X$. That is, for any $\phi \in Y^*$, the functional $< \phi, F(C + \lambda \tilde{C}) >$ defined on Y is an entire scalar valued function of λ. In Gil' (2008) (see also Lemma 2.14.1 from Gil' (2010a)), the following result has been proved.

Lemma 2.7 *Let $F(C + \lambda \tilde{C})$ $(\lambda \in \mathbf{C})$ be an entire function for all $C, \tilde{C} \in X$ and there be a monotone non-decreasing function $G : [0, \infty) \to [0, \infty)$, such that*

$$\|F(C)\|_Y \le G(\|C\|_X) \quad (C \in X). \tag{10.5}$$

Then

$$\|F(C) - F(\tilde{C})\|_Y \le \|C - \tilde{C}\|_X \; G(1 + \frac{1}{2}\|C + \tilde{C}\|_X + \frac{1}{2}\|C - \tilde{C}\|_X) \quad (C, \tilde{C} \in X).$$

Take $C = A$, $\tilde{C} = B$ and $F(A + \lambda B) = det\ (A + \lambda B)$. Then due to (10.2) we have (10.5) with $G(\|A\|) = \alpha_n \|A\|^n$. Now the previous lemma implies.

$$|det\ A - det\ B| \le \alpha_n |||A - B|||(1 + \frac{1}{2}|||A - B||| + \frac{1}{2}|||A + B|||)^n. \tag{10.6}$$

For a constant $c > 0$ put $A_1 = cA$ and $B_1 = cB$. Then by (10.6)

$$|det\ A_1 - det\ B_1| \le |||A_1 - B_1|||\alpha_n[1 + \frac{1}{2}|||A_1 - B_1||| + \frac{1}{2}|||A_1 + B_1|||]^n.$$

But

$$|||A_1||| = c|||A|||,\ |||B_1||| = c|||B||| \text{ and } |det\ A_1 - det\ B_1| = c^n |det\ A - det\ B|.$$

Thus,

$$c^n|det\ A - det\ B| \le c|||A - B|||(1 + cb)^n, \tag{10.7}$$

where

$$b = \frac{1}{2}|||A - B||| + \frac{1}{2}|||A + B|||.$$

Denote $x = bc$. Then from inequality (10.7) we obtain

$$|det\ A - det\ B| \le \alpha_n b^{n-1}|||A - B|||\frac{(1+x)^n}{x^{n-1}}.$$

Minimize the function

$$f(x) = \frac{(1+x)^n}{x^{n-1}},\ x > 0.$$

Simple calculation show that

$$\min_{x\ge 0} f(x) = \frac{n^n}{(n-1)^{n-1}}.$$

So

$$|det\ A - det\ B| \le \frac{\alpha_n n^n}{(n-1)^{n-1}} b^{n-1}|||A - B|||.$$

This is the assertion of the theorem. □

The previous theorem gives us two-sided inequalities for the determinants of matrices which are "close" to triangular ones. Indeed, denote by $\tilde{D}$, V_+ and V_- the diagonal, strictly upper triangular and strictly lower triangular parts, respectively, of a matrix $A = (a_{jk})_{j,k=1}^n$. Using the notation $A_+ = \tilde{D} + V_+$, we have

$$det\ A_+ = \prod_{k=1}^n a_{jj},$$

since A_+ is triangular. In addition, $A - A_+ = V_-$. Put $\delta(A) := |||V_-|||$. For example,

$$N_2^2(V_-) = \sum_{j=2}^n \sum_{k=1}^{j-1} |a_{jk}|^2.$$

For the norm

$$\|A\|_\infty =: \max_j \sum_{k=1}^{n} |a_{jk}|,$$

one has

$$\|V_-\|_\infty = \max_{j=2,\dots,n} \sum_{k=1}^{j-1} |a_{jk}|.$$

Now Theorem 2.11 implies

Corollary 2.14 *For a matrix $A = (a_{jk})_{j,k=1}^{n}$ we have*

$$\left| det\ A - \prod_{k=1}^{n} a_{jj} \right| \le \Delta(A),$$

where

$$\Delta(A) := \gamma_n \delta(A)\ (\delta(A) + \||A + A_+|\|)^{n-1},$$

and therefore,

$$\prod_{j=1}^{n}(|a_{jj}| - \Delta(A)) \le |det\ A| \le \prod_{j=1}^{n}(|a_{jj}| + \Delta(A)).$$

Consequently, A is invertible, provided

$$\prod_{k=1}^{n} |a_{jj}| > \Delta(A).$$

Clearly, one can exchange the places of V_+ and V_-. The latter result is sharp: if A is a triangular matrix, then we get the equalities, since $V_- = 0$ in this case.

Now let us recall the lower bound for determinants with dominant principal diagonals established by Ostrowski (1952).

Theorem 2.12 *Let $A = (a_{jk})$ be an $n \times n$-matrix and*

$$|a_{jj}| > \sum_{m=1, m\ne j} |a_{jm}| \quad (j = 1, \dots, n). \tag{10.8}$$

Then

$$|det\ A| \ge \prod_{j=1}^{n}(|a_{jj}| - \sum_{m=1, m\ne j}^{n} |a_{jm}|). \tag{10.9}$$

This theorem and Theorem 2.3 imply.

Corollary 2.15 *Let condition* (10.8) *hold. Then* $A \in \mathbf{C}^{n \times n}$ *is invertible and*

$$\|A^{-1}\| \leq \frac{N_2^{n-1}(A)}{(n-1)^{(n-1)/2} \prod_{j=1}^{n} (|a_{jj}| - \sum_{m=1, m \neq j}^{n} |a_{jm}|)}.$$

2.11 Bounds for the Eigenvalues

2.11.1 Gerschgorin's Circle Theorem

Let A be a complex $n \times n$-matrix, with entries a_{jk}. For $j = 1, \ldots, n$ write

$$R_j = \sum_{k=1}^{n} |a_{jk}|.$$

Let $\Omega(b, r)$ be the closed disc centered at $b \in \mathbf{C}$ with a radius r.

Theorem 2.13 (Gerschgorin) Every eigenvalue of A lies within at least one of the discs $\Omega(a_{jj}, R_j)$.

Proof Let λ be an eigenvalue of A and let $x = (x_j)$ be the corresponding eigenvector. Let i be chosen so that $|x_i| = max_j |x_j|$. Then $|x_i| > 0$, otherwise $x = 0$. Since x is an eigenvector, $Ax = \lambda x$ or equivalent

$$\sum_{k=1}^{n} a_{ik} = \lambda x_i$$

so, splitting the sum, we get

$$\sum_{k=1}^{n} a_{ik} x_k = \lambda x_i - a_{ii} x_i.$$

We may then divide both sides by x_i (choosing i as we explained we can be sure that $x_i \neq 0$) and take the absolute value to obtain

$$|\lambda - a_{ii}| \leq \sum_{k=1}^{n} |a_{ik}| \frac{|x_k|}{|x_i|} \leq R_i,$$

where the last inequality is valid because

$$\frac{|x_k|}{|x_l|} \leq 1.$$

As claimed. □

Note that for a diagonal matrix the Gerschgorin discs $\Omega(a_{jj}, R_j)$ coincide with the spectrum. Conversely, if the Gerschgorin discs coincide with the spectrum, the matrix is diagonal.

Corollary 2.16 *The spectral radius of a complex in general $n \times n$-matrix $A = (a_{jk})$ is subject to the inequality*

$$r_s(A) \leq \max_j \sum_{k=1}^{n} |a_{jk}|.$$

Moreover its lower spectral radius is subject to the inequality

$$r_l(A) \geq \min_j (|a_{jj}| - \sum_{k=1}^{n} |a_{jk}|),$$

provided

$$|a_{jj}| > \sum_{k=1}^{n} |a_{jk}| \ \ (j = 1, \ldots, n).$$

In addition,

$$\alpha(A) = \max_k Re\lambda_k(A) \leq \max_j Re\ a_{jj} + \sum_{k=1}^{n} |a_{jk}|$$

and

$$\beta(A) = \min_k Re\lambda_k(A) \geq \min_j Re\ a_{jj} - \sum_{k=1}^{n} |a_{jk}|.$$

2.11.2 Cassini Ovals

Brauer and Ostrovski independently observed that *each eigenvalue of an $n \times n$ matrix $A = (a_{jk})$, $n \geq 2$, is contained in the Cassini ovals*

$$\Omega(m, j) := \{\lambda \in \mathbf{C} : \ |\lambda - a_{mm}||\lambda - a_{jj}| \leq T_{mj}\}$$

where

$$T_{mj} := \left(\sum_{k=1}^{n} |a_{mk}|\right)\left(\sum_{k=1}^{n} |a_{jk}|\right)(j, m = 1, \ldots, n;\ j \neq m)$$

(see Sect. 3.2 from the survey Marcus and Minc (1964)).

That result leads to a better but more complicated localization of the spectrum of a matrix than Gerschgorin's theorem, cf. Varga (2004) and references therein.

From the Cassini ovals, for the eigenvalue λ satisfying $|\lambda| = r_s(A)$ there are indexes m and j such that eigenvalue λ it follows that

$$(r_s(A) - |a_{mm}|)(r_s(A) - |a_{jj}|) \leq |\lambda - a_{mm}||\lambda - a_{jj}| \leq T_{mj}.$$

Solving this inequality, we obtain that $r_s(A) \leq x_{jm}$, where

$$x_{jm} = \frac{1}{2}(|a_{mm}| + |a_{jj}|) + \sqrt{\frac{1}{4}(|a_{mm}| + |a_{jj}|)^2 + T_{mj} - |a_{mm}a_{jj}|}.$$

That is, x_{jm} is the larger zero of the polynomial

$$(x - |a_{mm}|)(x - |a_{jj}|) - T_{mj}.$$

We thus arrive at

Corollary 2.17 *The spectral radius of $n \times n$-matrix $A = (a_{jk})$ is subject to the inequality*

$$r_s(A) \leq \max_{m,j=1,\ldots,n;\ m\neq j} \left[\frac{1}{2}(|a_{mm}| + |a_{jj}|) + \sqrt{\frac{1}{4}(|a_{mm}| - |a_{jj}|)^2 + T_{mj}}\right].$$

2.11.3 The Perron Theorems

Again $A = (a_{jk})$ is a complex $n \times n$-matrix and for $j = 1, \ldots, n$ We write

$$R_j = \sum_{k=1}^{n} |a_{jk}|.$$

Recall that $\Omega(b, r)$ is the closed disc centered at $b \in \mathbf{C}$ with radius r.

Another valuable result is the following Perron theorem.

Theorem 2.14 *If $A = (a_{jk})$ is a non-negative matrix, then the spectral radius $r_s(A) = \max_k |\lambda_k(A)|$ of A is its eigenvalue.*

For the proof of this theorem see the book (Gantmakher 1967, Sect. 2, Chap. 13). We also state the following comparison theorem of Perron.

Theorem 2.15 *If a matrix* $A = (a_{jk})_{j,k=1}^n$ *is non-negative and if in matrix* $C = (c_{jk})_{j,k=1}^n$,

$$|c_{ji}| \le a_{ji} \quad (i, j = 1, 2, \dots n),$$

then any eigenvalue $\lambda(C)$ *of* C *satisfies the inequality* $|\lambda(C)| \le r_s(A)$.

Proof By the Gel'fand formula

$$r_s(A) = \lim_{n\to\infty} \sqrt[n]{\|A^n\|},$$

where the norm is assumed to be Euclidean. But clearly, $\|C^n\| \le \|A^n\|$. So $r_s(C) \le r_s(A)$. This proves the result. □

2.11.4 Bounds for the Eigenvalues of Matrices "Close" to Triangular Ones

In the present section we improve the previous bounds for the eigenvalues in the case when the matrix $A = (a_{jk})$ is "close" to a triangular one.

To this end let $\tilde{D}$, V_+ and V_- be the diagonal, upper nilpotent part and lower nilpotent part of matrix A, respectively. Using the notation $A_+ = \tilde{D} + V_+$, we arrive at the relations

$$g(A_+) = N_2(V_+) \text{ and } \|A - A_+\| = \|V_-\|.$$

Taking

$$\delta_A := \begin{cases} 2\|V_-\| & \text{if } \|V_-\| \ge N_2(V_+), \\ 2N_2^{1-1/n}(V_+)[\|V_-\|]^{1/n} & \text{if } \|V_-\| \le N_2(V_+), \end{cases}$$

due to Corollary 2.3 we obtain: $sv_{A_+}(A) \le \delta_A$. But A_+ is a triangular matrix. So its diagonal entries are the eigenvalues. We therefore arrive at the following result.

Corollary 2.18 *Let* $A = (a_{jk})_{j,k=1}^n$ *be an* $n \times n$*-matrix. Then for any eigenvalue* μ *of* A*, there is an index* $k = 1, \dots, n$*, such that*

$$|\mu - a_{kk}| \le \delta_A,$$

and therefore the (upper) spectral radius satisfies the inequality

$$r_s(A) \le \max_{k=1,\dots,n} |a_{kk}| + \delta_A,$$

and the lower spectral radius satisfies the inequality

$$r_l(A) \geq \min_{k=1,\dots,n} |a_{kk}| - \delta_A,$$

provided $|a_{kk}| > \delta_A$ $(k = 1, \dots, n)$.

Clearly, one can exchange the places of V_+ and V_-.

This result is sharp: if A is a triangular matrix, the we get the equalities $\lambda_k(A) = a_{kk}$.

In the case $a_{kk} \neq a_{jj}$ $(k \neq j)$ one can apply the results from Sect. 2.7.

2.12 Comments

Section 2.2. As it was above mentioned, the results contained in that section are well-known.

Sections 2.3 and 2.4 are based on Chap. 2 from Gil' (2003). Compare our results from Sect. 2.4 with the well-known ones. To this end put

$$w_n := \sum_{j=0}^{n-1} \frac{1}{\sqrt{j!}}.$$

Clearly $w_n < n$ $(n > 2)$. If

$$q w_n \leq g(A), \tag{12.1}$$

then for the positive root $\hat{z}(q/g(A))$ of Eq. (4.6), Lemma 2.2 gives us the inequality

$$\hat{z}(q/g(A)) \leq \sqrt[n]{q w_n / g(A)}.$$

Since $z(q, A) = \hat{z}(q/g(A))g(A)$, Theorem 2.4 yields the inequality

$$sv_A(\tilde{A}) \leq (q w_n)^{1/n} g^{1-1/n}(A). \tag{12.2}$$

The best known result of kind (12.2) or (4.7) is due to the Henrici theorem (see Stewart and Sun Ji-guang (1990, p. 172)). To compare our results with Henrici's recall that for a given matrix A we can find a unitary matrix U such that $UAU^* = D + V$, where D and V are the diagonal and the nilpotent parts of A (see Sect. 2.1). Such a V is not unique. If ν is any matrix norm, define the ν-departure from normality of A by $\delta_\nu(A) = inf \ \ \nu(V)$, where the infimum is taken over all V, which appear as the nilpotent parts in the upper triangular forms of A. For the Frobenius norm we have $\delta_F(A) = g(A)$, cf. Stewart and Sun Ji-guang (1990, p. 172).

Theorem 2.16 (Henrici) Let ν be a matrix norm , such that $\nu(C) \geq \|C\|$ for all $n \times n$ -matrices C. Then for every eigenvalue $\tilde{\lambda}$ of $\tilde{A}$ there is an eigenvalue λ of A, such that

$$\frac{(|\tilde{\lambda}-\lambda|/\delta_\nu(A))^n}{1+(|\tilde{\lambda}-\lambda|/\delta_\nu(A))+\cdots+(|\tilde{\lambda}-\lambda|/\delta_\nu(A))^{n-1}} \leq \frac{q}{\delta_\nu(A)}, \text{ where } q=\|\tilde{A}-A\|.$$

For a non-Frobenius norm the calculations of the departure from normality is a difficult task. In the case of the Frobenius norm the latter theorem it gives us the following result, cf. Stewart and Sun Ji-guang (1990, p. 173).

Corollary 2.19 *If*

$$\frac{q}{g(A)} < 1/n, \tag{12.3}$$

then

$$sv_A(\tilde{A}) \leq (qn)^{1/n} g^{1-1/n}(A). \tag{12.4}$$

If

$$\frac{q}{g(A)} \geq 1, \textit{ then } sv_A(\tilde{A}) \leq q+g(A).$$

For instance, let the condition (12.3) hold. Since $w_n < n$ $(n > 2)$ we can assert that our inequality (12.2) improves (12.4).

Let us recall the (Elsner) inequality:

$$sv_A(B) \leq q^{1/n}(\|A\|_n + \|B\|_n)^{1-1/n}, \tag{12.5}$$

cf. Stewart and Sun Ji-guang (1990, p. 168). This inequality holds also for so called the Hausdorff distance between the spectra of A and B. Assume that A is a normal matrix, then Theorem 2.4 gives us the equality $sv_A(B) = q$. So our result considerably improves (12.5) at least for matrices "close" to normal ones. For more information about perturbations of matrices, we refer the reader to the books Stewart and Sun Ji-guang (1990, p. 168) and Bhatia (2007).

The material of Sect. 2.5 is taken from Gil' (2013e). One of the first estimates for the norm of a regular matrix-valued function was established by Gel'fand and Shilov (1958) in connection with their investigations of partial differential equations, but that estimate is not sharp; it is not attained for any matrix. The problem of obtaining a sharp estimate for the norm of a matrix-valued function has been repeatedly discussed in the literature. In the late 1970s, the author has obtained a sharp estimate for a matrix-valued function regular on the convex hull of the spectrum. It is attained in the case of normal matrices. Later, that estimate was extended to various classes of nonselfadjoint operators, such as Hilbert-Schmidt operators, quasi-Hermitian operators (i.e., linear operators with completely continuous imaginary components), quasiunitary operators (i.e., operators represented as a sum of a unitary operator and a compact one), etc. Moreover, estimates for the norm of regular functions of infinite matrices, representing bounded operators in a Banach space with a Schauder basis also have been derived. For more details see Gil' (2003) and (2013e).

Sections 2.6–2.10 contain the results adopted from Gil' (2013e).

Section 2.11 contains mainly classical theorems, except Corollary 2.18, which is probably new.

For more information about the inequalities for the eigenvalues of matrices, see for instance (Marcus and Minc 1964). Many interesting results connected with Gerschgorin's circle theorem (Theorem 1.11.1) can be found in Varga (2004).

Chapter 3
Difference Equations with Continuous Time

The present chapter deals with linear autonomous and time-variant vector difference-delay equations with continuous time. Under various assumptions, norm estimates for the Cauchy (resolvent) operators of the considered equations are derived. In the sequel chapters these estimates give us stability conditions for the neutral type functional differential equations.

In this chapter, for a $p \geq 1$, $-\infty < a \leq 0$ and $0 < T \leq \infty$ for the simplicity we write $L^p(a, T)$ instead of $L^p([a, T], \mathbf{C}^n)$.

3.1 Autonomous Difference-Delay Equations

Let $R_0(\tau) = (r_{ij}(\tau))_{i,j=1}^n$ be a real $n \times n$-matrix-valued function defined on a finite segment $[0, \eta]$, whose entries have bounded variations $var\ (r_{ij})$ and finite numbers of jumps, and do not have jumps at zero. Consider the autonomous non-homogeneous difference-delay equation

$$w(t) - \int_0^\eta dR_0(\tau)w(t-\tau) = f(t) \ \ (t > 0), \tag{1.1}$$

with the zero initial function

$$w(t) = 0 \ (-\eta \leq t \leq 0). \tag{1.2}$$

Here f is a given function satisfying the conditions pointed below. The integral is understood as in Sect. 1.8. According to (1.2) for $t < \eta$ we have

$$\int_0^\eta dR_0(\tau)w(t-\tau) = \int_0^t dR_0(\tau)w(t-\tau).$$

M.I. Gil', *Stability of Neutral Functional Differential Equations*,
Atlantis Studies in Differential Equations 3, DOI 10.2991/978-94-6239-091-1_3

Thus (1.1) can be rewritten as

$$w - S_0 w = f, \tag{1.3}$$

where

$$S_0 w(t) = \int_0^{\min\{t,\eta\}} dR_0(\tau) w(t-\tau) \ \ (t > 0) \ \text{ and } \ S_0 w(t) \equiv 0 \ \ (t \le 0).$$

Example 3.1 Consider the equation

$$w(t) - Aw(t-h) = f(t) \ (0 < h < \eta, \ t > 0),$$

with condition (1.2) and an $n \times n$ constant matrix A. In the considered case

$$R_0(s) = \begin{cases} 0 & \text{if } \ 0 \le s \le h, \\ A & \text{if } \ h < s \le \eta. \end{cases}$$

In addition, S_0 can be defined by

$$S_0 w(t) = Aw(t-h) \ \ (t > h); \ S_0 w(t) = 0 \ \ (t \le h).$$

Example 3.2 Consider the equation

$$w(t) - \int_0^{\eta} A(s) w(t-s) ds = f(t) \ (t > 0),$$

with condition (1.2) and an $n \times n$ continuous matrix $A(s)$. In this case $dR_0(s) = A(s)ds$ and

$$S_0 w(t) = \int_0^{\min\{t,\eta\}} A(s) w(t-s) ds \ \ (t > 0) \ \text{ and } \ S_0 w(t) \equiv 0 \ \ (t \le 0).$$

Let f be a function measurable and bounded on any finite interval. Then a solution of problem (1.1), (1.2) is a function $w(t)$ measurable, bounded on any finite interval, satisfying (1.1) a.e. on $[0, \infty)$, and condition (1.2).

If for any function f measurable and bounded on $[0, \infty)$, problem (1.1), (1.2) has a solution w, then the operator $(I - S_0)^{-1}$ defined by

$$w = (I - S_0)^{-1} f$$

will be called *the Cauchy operator (the resolvent operator)* to that problem. If for some $p \ge 1$, $(I - S_0)^{-1}$ has a finite L^p-norm on the set of all bounded functions from $L^p(0,\infty)$, then it can be extended to the whole space $L^p(0,\infty)$.

Thanks to Corollary 1.2, there is a constant $V_{L^p}(R_0)$ independent of $T > 0$ and u, such that

$$\|\int_0^\eta dR_0(\tau)u(t-\tau)\|_{L^p(0,\infty)} \le V_{L^p}(R_0)\|u\|_{L^p(-\eta,\infty)} \; (u \in L^p(-\eta,\infty)).$$

Hence we easily have

$$\|S_0\|_{L^p(0,\infty)} \le V_{L^p}(R_0) \quad (p \ge 1). \tag{1.4}$$

Lemma 3.1 *Let the spectral radius $r_s(S_0; L^p)$ of operator S_0 in space $L^p(0,\infty)$ satisfy the inequality*

$$r_s(S_0; L^p) < 1. \tag{1.5}$$

Then

$$(I - S_0)^{-1} = \sum_{k=1}^{\infty} S_0^k. \tag{1.6}$$

Proof The series in (1.6) strongly converges and

$$(I - S_0)\sum_{k=1}^{\infty} S_0^k = \sum_{k=1}^{\infty} S_0^k(I - S_0) = I.$$

This proves the result. □

From this lemma we easily obtain that under condition (1.5),

$$\|(I - S_0)^{-1}\|_{L^p(0,\infty)} \le \sum_{k=1}^{\infty} \|S_0^k\|_{L^p(0,\infty)}. \tag{1.7}$$

Moreover, if

$$V_{L^p}(R_0) < 1, \tag{1.8}$$

then

$$\|(I - S_0)^{-1}\|_{L^p(0,\infty)} \le \frac{1}{1 - V_{L^p}(R_0)}. \tag{1.9}$$

Recall that

$$var(R_0) = \|(var\,(r_{jk}))^n_{j,k=1}\|_n,$$

i.e. $var(R_0)$ is the spectral (operator) norm of the matrix

$$\left(var\, r_{jk}\right)^n_{j,k=1}.$$

As it is shown in Sect. 1.8, the following inequalities are valid:

$$V_{L^2}(R_0) \le var\,(R_0),\ V_{L^\infty}(R_0) \le \sqrt{n}\, var\,(R_0).$$

In addition,

$$V_{L^p}(R_0) \le \sqrt{n}\, var\,(R_0) \quad \text{if} \quad p \ge 2,$$

and

$$V_{L^p}(R_0) \le \varsigma\,(R_0) \quad \text{if} \quad 1 \le p < 2,$$

where

$$\varsigma(R_0) = \sum_{j=1}^n \sqrt{\sum_{k=1}^n (var(r_{jk}))^2}.$$

Furthermore, in Sect. 1.8, the norm

$$\|f\|_{\hat{L}^\infty([a,b],\mathbf{C}^n)} = \|f\|_{\hat{L}^\infty(a,b)} = [\sum_{k=1}^n ess \sup_{a\le t\le b} |f_k(t)|^2]^{1/2}$$

has been introduced. Besides,

$$\|f\|_{L^\infty(a,b)} \le \|f\|_{\hat{L}^\infty(a,b)} \le \sqrt{n}\|f\|_{L^\infty(a,b)} \quad (f \in L^\infty(a,b)). \tag{1.10}$$

Lemma 1.9 asserts that

$$\|S_0\|_{\hat{L}^\infty(0,\infty)} \le var(R_0). \tag{1.11}$$

Assume that

$$var\,(R_0) < 1. \tag{1.12}$$

Then according to (1.9),

$$\|(I - S_0)^{-1}\|_{\hat{L}^\infty(0,\infty)} \le \frac{1}{1 - var(R_0)}.$$

Therefore for a solution of (1.1), (1.2) we obtain

$$\|w\|_{L^\infty(0,\infty)} \le \|w\|_{\hat{L}^\infty(0,\infty)} \le \|(I - S_0)^{-1}\|_{\hat{L}^\infty(0,\infty)} \|f\|_{\hat{L}^\infty(0,\infty)}.$$

Hence, due to (1.10),

$$\|w\|_{L^\infty(0,\infty)} \le \frac{\sqrt{n}}{1 - var(R_0)} \|f\|_{L^\infty([0,\infty],\mathbf{C}^n)} \quad (f \in L^\infty(0,\infty)).$$

We thus have proved.

Theorem 3.1 *Let condition* (1.12) *hold. Then*

$$\|(I - S_0)^{-1}\|_{L^\infty(0,\infty)} \le \frac{\sqrt{n}}{1 - var(R_0)} \ .$$

Take into account that S_0 maps bounded functions into bounded ones. So, if

$$r_s(S_0; L^\infty) < 1, \tag{1.13}$$

then according to (1.7) we can assert that a solution to (1.1) is bounded. Now the previous theorem implies

Corollary 3.1 *Let condition* (1.12) *hold and f be bounded on* $[0, \infty)$. *Then a solution to problem* (1.1), (1.2) *is subject to the inequality*

$$\|w\|_{C(0,\infty)} = \sup_{t\ge 0} \|w(t)\| \le \frac{\sqrt{n}\|f\|_{C(0,\infty)}}{1 - var(R_0)}. \tag{1.14}$$

Furthermore let f be defined on $(-\eta, \infty)$ and

$$E_0 f(t) = \int_0^\eta dR_0(\tau) f(t-\tau) \quad (t \ge 0).$$

Then $E_0 f(t)$ is continuous, if $f(t)$ is continuous. If $f(t) \equiv 0$ for $t \le 0$, then $f(t)$ is continuous on $(-\eta, \infty)$, provided it is continuous on $(0, \infty)$ and $f(0) = 0$.

Hence we deduce that $S_0 f$ is continuous if f is continuous and $f(0) = 0$. Since $S_0 f(0) = 0$ according to the definition of S_0, we can assert that $S_0^m f$ is continuous for any natural m. Thus we arrive at

Corollary 3.2 *Let condition* (1.12) *hold, f be continuous, bounded on* $[0, \infty)$ *and* $f(0) = 0$. *Then a solution to problem* (1.1), (1.2) *is continuous and is subject to inequality* (1.14).

3.2 Application of the Laplace Transform

It is supposed that Eq. (1.1) admits the Laplace transform. For example, if f is bounded, then under (1.13) a solution to (1.1), (1.2) is bounded and one can apply the Laplace transform. It gives us the equation

$$(I - \int_0^\eta e^{-zs} dR_0(s))\hat{w}(z) = \hat{f}(z) \quad (z \in \mathbf{C}),$$

where $\hat{w}(z)$ and $\hat{f}(z)$ are the Laplace transforms to w and f, respectively, z is the dual variable, and I is the unit matrix. Denote

$$H(\tau) = I - \int_0^\eta \tau^{-s} dR_0(s).$$

Then

$$H(e^z)\hat{w}(z) = \hat{f}(z) \quad (z \in \mathbf{C}),$$

$H(\tau)$ is called *the characteristic matrix-valued function* to Eq. (1.1). Besides *det* $H(\tau)$ is called *the characteristic determinant* of Eq. (1.1). A zero of the characteristic determinant *det* $H(z)$ is called *a characteristic value of* $H(.)$ and $\lambda \in \mathbf{C}$ is *a regular value* of $H(.)$ if *det* $H(\lambda) \neq 0$. The collection of all the characteristic values of H with their multiplicities is called *the spectrum of* H *and is denoted by* $\Sigma(H)$.

In the sequel it is assumed that all the characteristic values of $H(z)$ *are inside the disk* $|z| < 1$.

Below we also give some conditions that provide this property.

Furthermore, we have

$$\hat{w}(i\omega) = H^{-1}(e^{i\omega})\hat{f}(i\omega) \quad (\omega \in \mathbf{R}).$$

Hence,

$$\begin{aligned}\hat{\|}\hat{w}(i\omega)\|_{L^2(-\infty,\infty)} &= \|H^{-1}(e^{i\omega})\hat{f}(i\omega)\|_{L^2(0,\infty)} \\ &\leq \sup_{|z|=1} \|H^{-1}(z)\|_n \|\hat{f}(i\omega)\|_{L^2(-\infty,\infty)}.\end{aligned}$$

Making use the Parseval equality, see Sect. 1.10, we obtain

$$\|w\|_{L^2(0,\infty)} \leq \sup_{|z|=1} \|H^{-1}(z)\|_n \|f\|_{L^2(0,\infty)}.$$

We thus arrive at

Lemma 3.2 *If all the characteristic values of* $H(.)$ *are inside the disk* $|z| < 1$, *then*

$$\|(I - S_0)^{-1}\|_{L^2(0,\infty)} \leq \sup_{|z|=1} \|H^{-1}(z)\|_n.$$

3.3 L^p-Norms of Solutions to Autonomous Equations

3.3.1 Equations with One Delay in L^p, $p \geq 1$

Consider the equation

$$w(t) - Aw(t-h) = f(t) \ (t > 0), \tag{3.1}$$

with an $n \times n$-constant matrix A and a positive constant h and $f \in L^p(0, \infty)$, $p \geq 1$. Take the zero initial condition

$$w(t) = 0 \ (-h \leq t \leq 0). \tag{3.2}$$

It is assumed that the spectral radius $r_s(A)$ of A is less than one.

According to Example 3.1, we can write

$$R_0(s) = \begin{cases} 0 & \text{if } \ 0 \leq s \leq h, \\ A & \text{if } \ s > h. \end{cases}$$

and

$$S_0 w(t) = \begin{cases} Aw(t-h) & \text{if } \ t > h, \\ 0 & \text{if } \ t \leq h. \end{cases}$$

Hence,

$$S_0^k w(t) = \begin{cases} A^k w(t-kh) & \text{if } \ t > kh, \\ 0 & \text{if } \ t \leq kh. \end{cases}$$
$$(k = 1, 2, \ldots)$$

Consequently,

$$r_s(S_0, L^p) \leq r_s(A). \tag{3.3}$$

To estimate $r_s(A)$ one can apply the results from Sect. 2.11. In addition,

$$\|(I - S_0)^{-1}\|_{L^p(0,\infty)} \leq \sum_{m=0}^{\infty} \|A^m\|_n.$$

Recall that due to Corollary 2.5,

$$\|A^m\|_n \le \sum_{k=0}^{n-1} \frac{m! r_s^{m-k}(A) g^k(A)}{(m-k)!(k!)^{3/2}}.$$

Besides, $1/(m-k)! = 0$ if $m < k$. According to (1.7), we arrive at the inequality

$$\|(I - S_0)^{-1}\|_{L^p(0,\infty)} \le \sum_{k=0}^{n-1} \sum_{m=k}^{\infty} \frac{m! r_s^{m-k}(A) g^k(A)}{(m-k)!(k!)^{3/2}} \quad (p \ge 1).$$

But

$$\begin{aligned}\sum_{m=k}^{\infty} \frac{m! x^{m-k}}{(m-k)!} &= \frac{d^k}{dx^k} \sum_{m=0}^{\infty} x^m \\ &= \frac{d^k}{dx^k} \frac{1}{1-x} = \frac{k!}{(1-x)^{k+1}} \quad (0 < x < 1).\end{aligned}$$

We thus have proved the following result.

Lemma 3.3 *Let $r_s(A) < 1$. Then the Cauchy operator of* (3.1) *satisfies the inequality*

$$\|(I - S_0)^{-1}\|_{L^p(0,\infty)} \le \sum_{k=0}^{n-1} \frac{g^k(A)}{(1 - r_s(A))^{k+1}\sqrt{k!}} \quad (p \ge 1). \tag{3.4}$$

Since

$$g(A) \le \sqrt{2} N_2(A_I) \ \ (A_I = \frac{1}{2i}(A - A^*)),$$

where $N_2(A_I)$ is the Frobenius norm (see Sect. 2.1), we obtain

$$\|(I - S_0)^{-1}\|_{L^p(0,\infty)} \le \sum_{k=0}^{n-1} \frac{(\sqrt{2} N_2(A_I))^k}{(1 - r_s(A))^{k+1}\sqrt{k!}} \quad (p \ge 1).$$

Moreover, since

$$g(A) \le \sqrt{N_2^2(A) - Trace\ A^2},$$

(see Sect. 2.1), we can write

$$\|(I - S_0)^{-1}\|_{L^p(0,\infty)} \le \sum_{k=0}^{n-1} \frac{\left(\sqrt{N_2^2(A) - Trace\ A^2}\right)^k}{(1 - r_s(A))^{k+1}\sqrt{k!}} \quad (p \ge 1).$$

If A is diagonalizable, then Lemma 2.6 implies the inequality

$$\|A^m\|_n \le \kappa(A) r_s^m(A) \ (m = 1, 2, \ldots),$$

where $\kappa(A)$ is the condition number. Consequently,

$$\|(I - S_0)^{-1}\|_{L^p(0,\infty)} \le \kappa(A) \sum_{m=0}^{\infty} r_s^m(A)$$

and therefore the following result is true.

Lemma 3.4 *Let matrix A be diagonalizable and* $r_s(A) < 1$. *Then the Cauchy operator of* (3.1) *satisfies the inequality*

$$\|(I - S_0)^{-1}\|_{L^p(0,\infty)} \le \frac{\kappa(A)}{1 - r_s(A)}. \tag{3.5}$$

3.3.2 Autonomous Difference-Delay Equations with Several Delays

Introduce the space $M(a,\infty) = M([a,\infty), \mathbf{C}^n)$ of functions measurable and bounded on $[a,\infty)$ with the sup-norm $\|.\|_{C(a,\infty)} = \|.\|_{M(a,\infty)}$. Consider the equation

$$w(t) - \sum_{k=1}^{\nu} A_k w(t - h_k) = f(t) \ \ (t > 0) \tag{3.6}$$

with the conditions

$$w(t) = 0, t \le 0, \tag{3.7}$$

$n \times n$-constant matrices A_k and positive constants

$$0 < h_1 < h_2 < \cdots < h_\nu.$$

Introduce the operators

$$S_{0k}u(t) = A_k u(t - h_k) \ \ (t > h_k) \ \ \text{and} \ \ S_{0k}u(t) = 0 \ \ (t \le h_k),$$

and

$$S_0 = \sum_{k=1}^{\nu} S_{0k}.$$

In the considered case one can take

$$R_0(s) = \begin{cases} 0 & \text{if } 0 \leq s < h_1, \\ A_1 & \text{if } h_1 \leq s < h_2, \\ A_1 + A_2 & \text{if } h_2 \leq s < h_3, \\ \dots & \dots\dots \\ \sum_{k=1}^{\nu} A_k & \text{if } s \geq h_\nu. \end{cases}$$

Denote by $|x|_s$ the vector whose coordinates in the standard basis are the absolute values of the coordinates of a vector $x \in \mathbf{C}^n$ in that basis, and by $|A|_s$ the matrix whose entries are the absolute values of the entries of a matrix A in the standard basis. Then, for a function $u(t) = (u_k(t))_{k=1}^n$ defined and bounded on $[-h_\nu, \infty)$, and $u(t) = 0$ $(t \leq 0)$, in the coordinate-wise sense we obtain

$$|S_{0k}u|_s \leq |A_k|_s |u(t - h_k)|_s.$$

Introduce in $M(0, \infty)$ the generalised norm $\alpha(u)$ of $u(t)$ by the vector

$$\alpha(u) = (\sup_{t \geq 0} |u_k(t)|)_{k=1}^n.$$

About the generalised norm see Sect. 1.7 of this book. Simple calculations show that

$$\alpha(A_k u(t - h_k)) \leq |A_k|_s \alpha(u).$$

Hence,

$$\alpha(S_0 u) \leq \sum_{k=1}^{\nu} |A_k|_s \alpha(u) = B\alpha(u),$$

where

$$B = \sum_{k=1}^{\nu} |A_k|_s.$$

Thus,

$$\alpha(S_0^m u) \leq B\alpha(S_0^{m-1} u) \leq \dots \leq B^m \alpha(u).$$

But

$$\|\alpha(u)\|_n^2 = \sum_{k=1}^{n} \sup_{t \geq 0} |u_k(t)|^2 \geq \sup_{t \geq 0} \sum_{k=1}^{n} |u_k(t)|^2 = \|u\|_{C(0,\infty)}^2$$

and

$$\|\alpha(u)\|_n^2 = \sum_{k=1}^{n} \sup_{t\geq 0} |u_k(t)|^2 \leq n \sup_{t\geq 0, k=1,\ldots,n} |u_k(t)|^2 \leq n\|u\|^2_{C(0,\infty)}.$$

Consequently,

$$\|S_0^m u\|_{C(0,\infty)} \leq \|\alpha(S_0^m u)\|_n \leq \sqrt{n}\|B^m\|_n \|u\|_{C(0,\infty)}.$$

Hence it follows that the spectral radius $r_s(S_0, M)$ of S_0 in space $M(0,\infty)$ satisfies the inequality

$$r_s(S_0, M) \leq r_s(B)$$

and therefore

$$\|(I - S_0)^{-1}\|_{C(0,\infty)} \leq \sqrt{n} \sum_{m=0}^{\infty} \|B^m\|_n \quad (r_s(B) < 1). \tag{3.8}$$

According to (3.4) we obtain

$$\sum_{m=0}^{\infty} \|B^m\|_n \leq \sum_{k=0}^{n-1} \frac{g^k(B)}{(1 - r_s(B))^{k+1}\sqrt{k!}}$$

provided $r_s(B) < 1$. This proves our next result

Corollary 3.3 *Let* $B = |A_1|_s + \cdots + |A_\nu|_s$ *and* $r_s(B) < 1$. *Then the Cauchy operator of* (3.6) *satisfies the inequality*

$$\|(I - S_0)^{-1}\|_{C(0,\infty)} \leq \sqrt{n} \sum_{k=0}^{n-1} \frac{g^k(B)}{(1 - r_s(B))^{k+1}\sqrt{k!}}.$$

If B is diagonalizable, then due to Lemma 2.6 one can write

$$\|B^m\|_n \leq \kappa(B) r_s^m(B) \ (m = 1, 2, \ldots)$$

and therefore the (3.8) implies result.

Corollary 3.4 *Let* $B = |A_1|_s + \cdots + |A_\nu|_s$ *be diagonalizable and* $r_s(B) < 1$. *Then the Cauchy operator of* (3.6) *satisfies the inequality*

$$\|(I - S_0)^{-1}\|_{C(0,\infty)} \leq \frac{\kappa(B)\sqrt{n}}{1 - r_s(B)}. \tag{3.9}$$

3.4 L^2-Norms of Solutions to Autonomous Equations

In section we are going to improve the results of the previous section in the case $p = 2$.

Apply Corollary 2.1 to the characteristic matrix valued function $H(z)$ of Eq. (1.1) with a regular z. It gives us the inequality

$$\|H^{-1}(z)\|_n \leq \sum_{k=1}^{n-1} \frac{g^k(H(z))}{\sqrt{k!}d^{k+1}(H(z))}, \tag{4.1}$$

where

$$d(H(z)) := \min_k \left| 1 - \lambda_k \left(\int_0^\eta z^{-s} dR_0(s) \right) \right|.$$

But due to the property

$$g(zI + Ae^{it}) = g(A) \quad (t \in \mathbf{R}, z \in \mathbf{C}) \tag{4.2}$$

we get

$$g(H(z)) = g(\int_0^\eta z^{-s} dR_0(s)).$$

Assume that

$$var(R_0) = \int_0^\eta \|dR_0(s)\|_n < 1. \tag{4.3}$$

Since

$$|\lambda_k(\int_0^\eta z^{-s} dR_0(s))| \leq \int_0^\eta \|dR_0(s)\|_n = var(R_0) \quad (|z| = 1),$$

we get

$$d(H(z)) \geq 1 - var(R_0) \quad (|z| = 1),$$

and due to the definition of $g(A)$ we have $g(A) \leq N_2(A)$. Consequently, putting,

$$g_0(H) := \sup_{|z|=1} g\left(\int_0^\eta z^{-s} dR_0(s) \right),$$

we obtain

$$g_0(H) \le \sup_{|z|=1} N_2(\int_0^\eta z^{-s} dR_0(s)) \le \sqrt{n} \sup_{|z|=1} \| \int_0^\eta z^{-s} dR_0(s)\|_n.$$

Thus,

$$g_0(H) \le \sqrt{n}\, var(R_0). \tag{4.4}$$

Take into account that due Lemma 3.2,

$$\|(I - S_0)^{-1}\|_{L^2(0,\infty)} \le \sup_{|z|=1} \|H^{-1}(z)\|_n.$$

Now (4.1) implies

Lemma 3.5 *Let condition* (4.3) *hold. Then the Cauchy operator to Eq.* (1.1) *satisfies the inequalities*

$$\|(I - S_0)^{-1}\|_{L^2(0,\infty)} \le \sum_{k=1}^{n-1} \frac{g_0^k(H)}{\sqrt{k!}(1 - var(R_0))^{k+1}} \le \sum_{k=1}^{n-1} \frac{(\sqrt{n}\, var(R_0))^k}{\sqrt{k!}\,(1 - var(R_0))^{k+1}}. \tag{4.5}$$

In particular, consider the equation

$$w(t) - A\int_0^\eta w(t-s)d\mu(s) = f(t) \quad (t > 0) \tag{4.6}$$

with the zero initial conditions. Here A is an $n \times n$—matrix and μ is a nondecreasing function with a finite number of jumps.

The characteristic matrix valued function $H(z)$ of this equation is

$$H(z) = I - A\int_0^\eta z^{-s} d\mu(s).$$

Then by the definition of g and property (4.2),

$$g(H(z)) = g(A\int_0^\eta z^{-s} d\mu) \le g(A)\int_0^\eta d\mu = g(A) var(\mu) \quad (|z| = 1).$$

So $g_0(H) \le g(A)var(\mu)$. In addition,

$$d(H(z) \geq 1 - \max_k |\lambda_k(A) \int_0^\eta z^{-s} d\mu|$$

$$\geq 1 - \max_k |\lambda_k(A)| var(\mu) \quad (|z| = 1).$$

Hence, $d_0(H) \geq 1 - r_s(A)var(\mu)$, provided

$$r_s(A)var(\mu) < 1. \tag{4.7}$$

We thus arrive at our next result.

Lemma 3.6 *Let condition* (4.7) *hold. Then the Cauchy operator to Eq.* (4.6) *satisfies the inequality*

$$\|(I - S_0)^{-1}\|_{L^2(0,\infty)} \leq \sum_{k=1}^{n-1} \frac{(g(A)var(\mu))^k}{\sqrt{k!}(1 - r_s(A)\ var(\mu))^{k+1}}.$$

For instance, for the equation

$$w(t) - Aw(t - h) = f(t) \quad (t > 0), \tag{4.8}$$

one has

$$H(z) = I - Az^{-h} \tag{4.9}$$

and $var(\mu) = 1$. Condition (4.7) takes the form

$$r_s(A) < 1. \tag{4.10}$$

Recall that *a matrix A with the property* (4.10) *is called a Schur stable matrix.*

3.5 Solution Estimates Via Determinants

Due to Theorem 2.3, for any invertible $n \times n$-matrix A $(n \geq 2)$ one has

$$\|A^{-1}\ det\ A\| \leq \frac{N_2^{n-1}(A)}{(n-1)^{(n-1)/2}}$$

and therefore,

$$\|H^{-1}(z)\| \le \frac{N_2^{n-1}(H(z))}{(n-1)^{(n-1)/2}|det\ H(z)|}$$

for any regular z. But

$$N_2(H(z)) = N_2(I - \int_0^\eta z^{-s} dR_0(s))$$

$$\le \sqrt{n}(1 + \int_0^\eta \|dR_0(s)\|) = \sqrt{n}(1 + var\ (R_0))\ \ (|z| = 1).$$

So

$$\|H^{-1}(z)\| \le \left(\frac{n}{n-1}\right)^{(n-1)/2} \frac{(1 + var\ (R_0))^{n-1}}{|det\ H(z)|}. \tag{5.1}$$

The entries $w_{jk}(z)$ of $H(z)$ are

$$w_{jk}(z) = -\int_0^\eta z^{-s} dr_{jk}(s)\ \ (k \ne j),$$

and

$$w_{jj}(z) = 1 - \int_0^\eta z^{-s} dr_{jj}(s)\ \ (k, j = 1, \dots, n).$$

Let $|z| = 1$. If,

$$|1 - w_{jj}(z)| > \sum_{m=1, m\ne j}^{n} |w_{jm}(z)|,$$

then due to the Ostrowski determinant inequality (see Sect. 2.10), we obtain

$$|det\ H(z)| \ge \prod_{j=1}^{n} \left(|1 - w_{jj}(z)| - \sum_{m=1, m\ne j}^{n} |w_{jm}(z)| \right) > 0.$$

But $|w_{jk}(z)| \le var(r_{jk})\ \ (k \ne j)$ and, in addition,

$$|1 - w_{jj}(z)| > 1 - var(r_{jj}),$$

provided $var(r_{jj}) < 1$. Thus

$$|det\ H(z)| \geq \prod_{j=1}^{n}(1 - \sum_{m=1}^{n} var(r_{jm})) > 0 \ \ (|z| = 1),$$

provided

$$\sum_{m=1}^{n} var(r_{jm}) < 1 \ \ (j = 1, \ldots, n). \tag{5.2}$$

Under this condition, due to (5.1) we have

$$\|H^{-1}(z)\|_n \leq \tau(R_0) \ \ (|z| = 1),$$

where

$$\tau(R_0) := \left(\frac{n}{n-1}\right)^{(n-1)/2} \frac{(1 + var\ (R_0))^{n-1}}{\prod_{j=1}^{n}(1 - \sum_{m=1}^{n} var(r_{jm}))}.$$

Now Lemma 3.2 implies

Lemma 3.7 *Let condition* (5.2) *hold. Then the Cauchy operator to* (1.1) *satisfies the inequality*

$$\|(I - S_0)^{-1}\|_{L^2(0,\infty)} \leq \tau(R_0).$$

For example, if $A = (a_{jk})_{j,k=1}^{n}$ and

$$H(z) = I - Az^{-h}, \tag{5.3}$$

then condition (5.2) takes the form

$$\sum_{m=1}^{n} |a_{jm}| < 1 \ (j = 1, \ldots, n), \tag{5.4}$$

and thus under condition (5.4), according to (5.1), the Cauchy operator to (4.8) satisfies the inequalities

$$\begin{aligned}\|(I - S_0)^{-1}\|_{L^2(0,\infty)} &\leq \left(\frac{1}{n-1}\right)^{(n-1)/2} \frac{(\sqrt{n} + N_2(A))^{n-1}}{\prod_{j=1}^{n}(1 - \sum_{m=1}^{n} |a_{jm}|)} \\ &\leq \left(\frac{n}{n-1}\right)^{(n-1)/2} \frac{(1 + \|A\|_n)^{n-1}}{\prod_{j=1}^{n}(1 - \sum_{m=1}^{n} |a_{jm}|)}.\end{aligned}$$

3.6 Time-Variant Equations: The General Case

In this section $R(t,\tau) = (r_{ij}(t,\tau))_{i,j=1}^n$ is a real $n \times n$-matrix-valued functions defined on $[0,\infty)\times[0,\eta]$ whose entries are continuous in t, have uniformly bounded variations in τ:

$$v_{jk} := \sup_{t\geq 0} var\, r_{ij}(t,.) < \infty \tag{6.1}$$

finite numbers of jumps in τ and do not have jumps at $\tau = 0$.

Let us consider the difference-delay equation

$$w(t) - \int_0^\eta d_\tau R(t,\tau)w(t-\tau) = f(t) \ \ (t > 0), \tag{6.2}$$

with the zero initial condition

$$w(t) = 0 \ (-\eta \leq t \leq 0). \tag{6.3}$$

and a given function $f(t)$. The integral is understood as in Sect. 1.9.

A solution of problem (6.2), (6.3) is defined as in the autonomous case (see Sect. 3.1).

Introduce at $u \in L^p(0,\infty)$ $(p \geq 1;\ T > 0)$ the operator S by

$$Su(t) = \int_0^{\min\{t,\eta\}} d_\tau R(t,\tau)u(t-\tau) \ \ (t > 0) \ \ \text{and} \ \ Su(t) \equiv 0 \ \ (t \leq 0). \tag{6.4}$$

The Cauchy operator $(I-S)^{-1}$ to Eq. (6.2), as in the autonomous case, is defined by $(I-S)^{-1}f = w$, where w is a solution of problem (6.2), (6.3).

Problem (6.2), (6.3) can be rewritten as

$$(I-S)w = f. \tag{6.5}$$

From Lemma 1.10 and its corollary it follows that for any $p \geq 1$ there is a constants $V_{L^p}(R)$ independent of $T > 0$ and $u \in L^p(-\eta,T)$, such that

$$\|\int_0^\eta d_\tau R(t,\tau)u(t-\tau)\|_{L^p(0,T)} \leq V_{L^p}(R)\|u\|_{L^p(-\eta,T)}.$$

Hence we easily obtain

$$\|S\|_{L^p(0,\infty)} \leq V_{L^p}(R).$$

We need the following result.

Lemma 3.8 *Let the spectral radius $r_s(S; L^p)$ of operator S in $L^p(0,\infty)$ $(p \ge 1)$ satisfy the inequality*

$$r_s(S; L^p) < 1. \tag{6.6}$$

Then

$$(I - S)^{-1} = \sum_{k=1}^{\infty} S^k.$$

The proof of this lemma is similar to the proof of Lemma 3.1.

From this result, it follows that under condition (6.6), one has

$$\|(I - S)^{-1}\|_{L^p(0,\infty)} \le \sum_{k=1}^{\infty} \|S^k\|_{L^p(0,\infty)} \quad (p \ge 1). \tag{6.7}$$

Moreover, if

$$V_{L^p}(R) < 1, \tag{6.8}$$

then

$$\|(I - S)^{-1}\|_{L^p(0,\infty)} \le \frac{1}{1 - V_{L^p}(R)} \,. \tag{6.9}$$

We need the quantity

$$V(R) := \|(v_{jk})\|_n \text{ (the spectral norm of matrix } (v_{jk})_{j,k=1}^n). \tag{6.10}$$

As it is shown in Sect. 1.9 the following inequalities are valid:

$$V_{L^2}(R) \le V(R), V_{L^\infty}(R) \le \sqrt{n}\, V(R).$$

In addition,

$$V_{L^p}(R) \le \sqrt{n}\, V(R) \quad \text{if} \quad p \ge 2,$$

and

$$V_{L^p}(R) \le \varsigma\,(R) := \sum_{j=1}^{n} \sqrt{\sum_{k=1}^{n} v_{jk}^2}, \quad \text{if} \quad 1 \le p < 2.$$

Furthermore, Lemma 1.11 implies the inequality

$$\|S\|_{\hat{L}^\infty(0,\infty)} \le V(R). \tag{6.11}$$

Assume that

$$V(R) < 1. \tag{6.12}$$

Then

$$\|(I-S)^{-1}\|_{\hat{L}^\infty(0,\infty)} \le \frac{1}{1-V(R)}.$$

Therefore, according to (1.10), for a solution of (6.2), (6.3) we obtain

$$\|w\|_{L^\infty(0,\infty)} \le \|w\|_{\hat{L}^\infty(0,\infty)} \le \|(I-S)^{-1}\|_{\hat{L}^\infty(0,\infty)} \|f\|_{\hat{L}^\infty(0,\infty)}.$$

Consequently,

$$\|w\|_{L^\infty(0,\infty)} \le \frac{\sqrt{n}}{1-V(R)} \|f\|_{L^\infty(0,\infty)} \quad (f \in L^\infty(0,\infty)). \tag{6.13}$$

We thus have proved.

Theorem 3.2 *Let condition* (6.12) *hold. Then*

$$\|(I-S)^{-1}\|_{L^\infty(0,\infty)} \le \frac{\sqrt{n}}{1-V(R)}.$$

Taking into account (6.12) we can write $\|S^m\|_{\hat{L}^\infty(0,\infty)} \le V^m(R)$. Since $\|.\|_{\hat{L}^\infty(0,\infty)}$ and $\|.\|_{L^\infty(0,\infty)}$ are equivalent norms, we have $\|S^m\|_{L^\infty(0,\infty)} \le const\ V^m(R)$. Thus under (6.12), the inequality $r_s(S, L^\infty) < 1$ is true. It is clear that that S maps $M(0,\infty)$ into itself. Consequently,

$$\|S^m f\|_{L^\infty(0,\infty)} = \|S^m f\|_{C(0,\infty)} \quad (f \in M(0,\infty)).$$

Thus we deduce that

$$\|w\|_{C(0,\infty)} \le \sum_{k=1}^{\infty} \|S^k f\|_{C(0,\infty)} \le \frac{const}{1-V(R)} \|f\|_{C(0,\infty)}.$$

So $w \in M(0,\infty)$ and (6.13) holds with $\|.\|_{C(0,\infty)}$ instead of $\|.\|_{L^\infty(0,\infty)}$. Making us the previous theorem, we arrive at

Corollary 3.5 *Let condition* (6.12) *hold. Then operator* $(I - S)^{-1}$ *is bounded in* $M(0, \infty)$ *and*

$$\|(I - S)^{-1}\|_{C(0,\infty)} \leq \frac{\sqrt{n}}{1 - V(R)}.$$

Consider now the homogeneous difference equation

$$y(t) - \int_0^\eta d_\tau R(t, \tau) y(t - \tau) = 0 \quad (t > 0) \tag{6.14}$$

with the initial condition

$$y(t) = \phi(t) \ (-\eta \leq t \leq 0) \tag{6.15}$$

for a given continuous function ϕ. A solution of this problem is defined as a function $y(t)$ bounded on each finite interval of the positive half-line and satisfying Eq. (6.14) and condition (6.15). Let

$$\hat{\phi}(t) = \phi(t) \ (-\eta \leq t \leq 0) \quad \text{and} \quad \hat{\phi}(t) = \phi(0) \ (t > 0),$$

and substitute $w(t) = y(t) - \hat{\phi}(t)$. Then we obtain problem (6.2), (6.3). So our above results for the problem (6.2), (6.3) give us the existence and uniqueness of solutions to problem (6.14), (6.15). The details are left to the reader.

Equation (6.14) is said to be *stable*, if for any $\phi \in C(-\eta, 0)$, a solution $y(t)$ of problem (6.14), (6.15) satisfies the inequality

$$\|y\|_{C(0,\infty)} \leq m_1 \|\phi\|_{C(-\eta,0)},$$

where m_1 is a constant independent on ϕ.

Equation (6.14) is said to be *exponentially stable*, if there are constants $m_0 \geq 1$ and $0 < \nu < 1$, such that a solution $y(t)$ of problem (6.14), (6.15) satisfies the inequality

$$\|y\|_{C(0,\infty)} \leq m_0 \nu^t \|\phi\|_{C(-\eta,0)} \ (t \geq 0)$$

for any $\phi \in C(-\eta, 0)$.

It is not hard to prove that the autonomous difference equation is exponentially stable if and only if all its characteristic values are in the open unit disk centered at zero.

3.7 Time-Variant Difference Equations with One Delay

Consider the equation

$$w(t) - A(t) w(t - h(t)) = f(t) \ (t > 0), \tag{7.1}$$

with the zero initial condition

$$w(t) = 0 \ (t \leq 0),$$

a variable $n \times n$-matrix $A(t)$ uniformly bounded and continuous on the positive half-line and a scalar continuous function $h(t)$ having the property

$$0 \leq h(t) < \eta < \infty \ (t \geq 0).$$

Put

$$Su(t) = A(t)u(t - h(t)) \ \ (t > h(t)); \ \ Su(t) = 0 \ (t \leq h(t)).$$

In the considered case

$$R(t, s) = \begin{cases} 0 & \text{if} \ \ 0 \leq s \leq h(t), \\ A(t) & \text{if} \ \ h(t) < s \leq \eta. \end{cases}$$

Let $|x|_s$ denote the vector whose coordinates are the absolute values of the coordinates of a vector x in the standard basis, and $|A(t)|_s$ be the matrix whose entries are the absolute values of the entries of a matrix $A(t)$ in the standard basis.

Assume that there is a positive constant matrix B, such that in the entry-wise sense

$$|A(t)|_s \leq B \ \ (t \geq 0). \tag{7.2}$$

Then

$$|Su(t)|_s \leq |A(t)|_s |u(t - h(t))|_s \leq B|u(t - h(t))|_s.$$

Again introduce the generalised norm $\alpha(u)$ of a bounded function $u(t) = (u_k(t))_{k=1}^n$ by the vector

$$\alpha(u) = (\sup_{t \geq 0} |u_k(t)|)_{k=1}^n.$$

Then we have

$$\alpha(Su) \leq B\alpha(u) \ \ (u \in M(0, \infty)).$$

Hence,

$$\alpha(S^m u) \leq B\alpha(S^{m-1}u) \leq \cdots \leq B^m \alpha(u).$$

Taking into account that

$$\|u\|_{C(0,\infty)} \leq \|\alpha(u)\|_n \leq \sqrt{n}\|u\|_{C(0,\infty)}$$

(see Sect. 3.3), we obtain

$$\|S^m u\|_{C(0,\infty)} \le \|\alpha(S^m u)\|_n \le \sqrt{n}\|B^m\|_n\|u\|_{C(0,\infty)}.$$

Hence it follows that

$$r_s(S, M) \le r_s(B).$$

So if the conditions (7.2) and

$$r_s(B) < 1 \tag{7.3}$$

hold, then the Cauchy operator to Eq. (7.1) satisfies the inequality

$$\|(I - S)^{-1}\|_{C(0,\infty)} \le \sqrt{n}\sum_{m=0}^{\infty} \|B^m\|_n.$$

To estimate $r_s(B)$ one can apply the results from Sect. 2.11.

Repeating arguments of the proofs of Lemmas 3.3 and 3.4 we arrive at

Theorem 3.3 *Let the conditions* (7.2) *and* (7.3) *hold. Then the Cauchy operator to Eq.* (7.1) *satisfies in space* $M(0, \infty)$ *the inequality*

$$\|(I - S)^{-1}\|_{C(0,\infty)} \le \sqrt{n}\sum_{k=0}^{n-1} \frac{g^k(B)}{(1 - r_s(B))^{k+1}\sqrt{k!}}.$$

If, in addition, B is diagonalizable, then

$$\|(I - S)^{-1}\|_{C(0,\infty)} \le \frac{\kappa(B)\sqrt{n}}{1 - r_s(B)}.$$

Recall that $\kappa(B)$ means the condition number (see Sect. 2.7).

3.8 Time-Variant Equations with Several Delays

Consider the problem

$$w(t) - \sum_{k=1}^{\nu} A_k(t)w(t - h_k(t)) = f(t) \;\; (t > 0), \tag{8.1}$$

$$w(t) = 0 \;\; (t \le 0)$$

with variable $n \times n$-matrices $A_k(t)$ uniformly bounded and continuous on the positive half-line and scalar continuous functions $h_k(t)$ satisfying

$$0 \le h_1(t) < h_2(t) < \cdots < h_\nu(t) < \eta.$$

Introduce operators S_k by

$$S_k u(t) = A_k(t) w(t - h_k(t)) \ \ (t > h_k(t)) \quad \text{and} \quad S_k u(t) = 0 \ (t \le h_k(t)),$$

and, in addition, put

$$Su(t) = \sum_{k=1}^{\nu} S_k.$$

In the considered case

$$R(s) = \begin{cases} 0 & \text{if } \ 0 \le s \le h_1(t), \\ A_1(t) & \text{if } \ h_1(t) < s \le h_2(t), \\ A_1(t) + A_2(t) & \text{if } \ h_2(t) < s \le h_3(t), \\ \cdots\cdot & \cdots\cdots\cdot \\ \sum_{k=1}^{\nu} A_k(t) & \text{if } \ s > h_\nu(t). \end{cases}$$

Recall that the symbol $|.|_s$ is introduced in Sect. 3.3. Assume that

$$|A_k(t)|_s \le B_k \ \ (k = 1, \ldots, \nu; \ t \ge 0) \tag{8.2}$$

for a positive constant matrix B_k. Then

$$|Su(t)|_s \le \sum_{k=1}^{\nu} B_k |u(t - h(t))|_s \ \ (u \in M(-\eta, \infty)).$$

Let $\alpha(u)$ be the generalised norm defined as in the previous section. Then

$$\alpha(Su) \le \sum_{k=1}^{s} B_k \alpha(u) = B_0 \alpha(u),$$

where

$$B_0 = \sum_{k=1}^{\nu} B_k. \tag{8.3}$$

Consequently,

$$\alpha(S^m u) \le B_0 \alpha(S^{m-1} u) \le \cdots \le B_0^m \alpha(u).$$

Taking into account that

$$\|u\|_{C(0,\infty)} \le \|\alpha(u)\|_n \le \sqrt{n}\|u\|_{C(0,\infty)}$$

(see Sect. 3.3), we obtain

$$\|S^m u\|_{C(0,\infty)} \le \sqrt{n}\|B_0^m\|_n \|u\|_{C(0,\infty)}.$$

Hence it follows that the spectral radius $r_s(S, M)$ of S in space $M(0, \infty)$ satisfies the inequality

$$r_s(S, M) \le r_s(B_0).$$

So, if

$$r_s(B_0) < 1, \tag{8.4}$$

then the Cauchy operator $(I - S)^{-1}$ to Eq. (8.1) satisfies the inequality

$$\|(I - S)^{-1}\|_{C(0,\infty)} \le \sqrt{n} \sum_{m=0}^{\infty} \|B_0^m\|_n.$$

Hence, according to Theorem 3.3 and Lemma 3.4 we get

Theorem 3.4 *Let conditions* (8.2) *hold,* B_0 *be defined by* (8.3), *and inequality* (8.4) *be fulfilled. Then the Cauchy operator to Eq.* (8.1) *in space* $M(0, \infty)$ *satisfies the inequality*

$$\|(I - S)^{-1}\|_{C(0,\infty)} \le \sqrt{n} \sum_{k=0}^{n-1} \frac{g^k(B_0)}{(1 - r_s(B_0))^{k+1}\sqrt{k!}}.$$

If, in addition, B_0 *is diagonalizable, then*

$$\|(I - S)^{-1}\|_{C(0,\infty)} \le \frac{\kappa(B_0)\sqrt{n}}{1 - r_s(B_0)}.$$

3.9 Difference Equations with Commensurable Delays

This section is devoted to the vector difference equation

$$\sum_{k=1}^{\mu} A_k w(t - k) = 0 \quad (t > 0;\ A_0 = I_n;\ \mu < \infty) \tag{9.1}$$

with constant matrix coefficients A_k. Here I_n means the unit $n \times n$-matrix

To Eq. (9.1) can be reduced the equation

$$\sum_{k=1}^{\mu} A_k w_1(s - hk) = 0$$

where $h \neq 1$ is a positive constant, if we set $t = s/h$ and $w(t) = w_1(th)$. The characteristic matrix function of Eq. (9.1) is

$$H(z) = \sum_{k=1}^{\mu} A_k z^{-k} = z^{-\mu} P(z),$$

where

$$P(z) = \sum_{k=1}^{\mu} A_{\mu-k} z^k.$$

Thus the characteristic values of $P(z)$ coincide with the characteristic values of $H(z)$, since $A_\mu \neq 0$.

Let $z_k(P)$ be the characteristic values of P taken with their multiplicities be ordered in the non-increasing way of their absolute values:

$$|z_k(P)| \geq |z_{k+1}(P)| \ \ (k = 1, \ldots, \mu n - 1).$$

As it is well-known (Rodman 1989), the spectrum $\Sigma(P)$ of the matrix polynomial P coincides with the spectrum $\sigma(A(P))$ of the $\mu \times \mu$-block matrix

$$A(P) := \begin{pmatrix} -A_1 & -A_2 & \ldots & -A_{\mu-1} & -A_\mu \\ I_n & 0 & \ldots & 0 & 0 \\ 0 & I_n & \ldots & 0 & 0 \\ . & . & \ldots & . & . \\ 0 & 0 & \ldots & I_n & 0 \end{pmatrix}.$$

Furthermore, introduce the $n \times n$-matrix

$$\Theta_P := \sum_{k=1}^{\mu} A_k A_k^*,$$

where the asterisk means the conjugate. Let $\lambda_k(\Theta_P)$ be the characteristic values of Θ_P taken with their multiplicities and enumerated in the non-increasing way: $|\lambda_k(\Theta_P)| \geq |\lambda_{k+1}(\Theta_P)|$. Denote

$$\omega_k(P) = \begin{cases} \sqrt{\lambda_k(\Theta_P)} & \text{if } \ k = 1, \ldots, n, \\ 0 & \text{if } \ k = n+1, \ldots, \mu n \end{cases}.$$

Theorem 3.5 *The characteristic values of P satisfy the inequalities*

$$\sum_{k=1}^{j} |z_k(P)| < j + \sum_{k=1}^{j} \omega_k(P) \quad (j = 1, \dots, \mu n).$$

For the proof see Theorem 11.1.2 from Gil' (2010a).

Let $\nu_P(r)$ denote the counting function of P. That is, $\nu_P(r)$ is the number of the characteristic values of P inside the disc $|z| \le r$.

Theorem 3.5 yields the following result.

Corollary 3.6 *The counting function $\nu_P(r)$ of the characteristic values of P satisfies the inequality*

$$\nu_P(r) \le \mu n - j + 1 \quad (j = 1, 2, \dots, \mu n)$$

for any

$$r \le 1 + \frac{1}{j} \sum_{k=1}^{j} \omega_k(P).$$

Let us point some identities for the sums

$$s_m(P) := \sum_{k=1}^{n\mu} z_k^m(P)$$

for an integer $m \le \mu$.

To formulate the result, introduce the $m \times m$-block matrix

$$\hat{B}_m = \begin{pmatrix} -A_1 & -A_2 & \dots & -A_{m-1} & -A_m \\ I_n & 0 & \dots & 0 & 0 \\ 0 & I_n & \dots & 0 & 0 \\ . & . & \dots & . & . \\ 0 & 0 & \dots & I_n & 0 \end{pmatrix} \quad (1 \le m \le \mu)$$

and

$$\hat{B}_1 = -A_1.$$

Clearly, $\hat{B}_\mu = A(P)$.

Theorem 3.6 *For any positive integer $m \le \mu$, one has*

$$\sum_{k=1}^{\mu n} z_k^m(P) = Trace\ \hat{B}_m^m.$$

For the proof see Theorem 11.2.1 from Gil' (2010a). Let us point also the following result proved in Gil' (2010a), Theorem 11.6.2.

Theorem 3.7 *Let*

$$m_k = \|A_k\|_n \ \ (k = 1, 2, \dots, \mu).$$

Then the spectrum of the pencil

$$P(\lambda) = \lambda^\mu + A_1\lambda^{\mu-1} + \cdots + A_\mu$$

lies in the set

$$\{z \in \mathbf{C} : |z| \le \tilde{r}\},$$

where $\tilde{r}$ is the unique positive root of the equation

$$z^\mu = \sum_{k=1}^{\mu} m_k z^{\mu-k}.$$

The matrix polynomial P is said to be *Schur stable*, provided all its values lie inside the unit disc $|z| < 1$.

Now Lemma 2.2, Corollary 2.2 and the previous theorem imply.

Corollary 3.7 *Let at least one of the conditions*

$$\sum_{k=1}^{\mu} \|A_k\|_n < 1$$

or

$$2 \max_{j=0,\dots,\mu n-1} \sqrt[j+1]{\|A_j\|_n} < 1$$

hold. Then P is Schur stable and therefore Eq. (9.1) *is asymptotically stable.*

3.10 Perturbations of Characteristic Values

Let

$$P(z) = \sum_{k=1}^{\mu} A_{\mu-k} z^k \ \text{ and } \ \tilde{P}(z) = \sum_{k=1}^{\mu} \tilde{A}_{\mu-k} z^k \ (z \in \mathbf{C}) \tag{10.1}$$

with constant $n \times n$-matrices $A_k, \tilde{A}_k$ $(k = 1, \dots, \mu)$, $A_0 = \tilde{A}_0 = I_n$. As it is shown in the previous section, the spectrum of $P(z)$ coincides with the spectrum of the characteristic function of (9.1). Similarly, $\Sigma(\tilde{P}(.))$ coincides with $\Sigma(\tilde{H}(.))$, where

$$\tilde{H}(z) = \sum_{k=1}^{\mu} \tilde{A}_k z^{-k} = z^{-\mu} \tilde{P}(z)$$

is the characteristic function of the difference equation

$$\sum_{k=1}^{\mu} \tilde{A}_k w(t-k) = 0 \quad (t > 0). \tag{10.2}$$

So the characteristic values of $\tilde{P}(z)$ coincide with the characteristic values of $\tilde{H}(z)$.

As it was mentioned in the previous section, the spectrum $\Sigma(P)$ of the matrix polynomial P coincides with the spectrum $\sigma(A(P))$ of the block matrix $A(P)$. Similarly, $\Sigma(\tilde{P})$ coincides with the spectrum $\sigma(A(\tilde{P}))$ of the $\mu \times \mu$ block matrix

$$A(\tilde{P}) := \begin{pmatrix} -\tilde{A}_1 & -\tilde{A}_2 & \dots & -\tilde{A}_{\mu-1} & -\tilde{A}_\mu \\ I_n & 0 & \dots & 0 & 0 \\ 0 & I_n & \dots & 0 & 0 \\ . & . & \dots & . & . \\ 0 & 0 & \dots & I_n & 0 \end{pmatrix}.$$

We are going to investigate the spectral variation of $A(\tilde{P})$ with respect to $A(P)$ with the aid of Theorem 2.4. To this end note that

$$g^2(A(P)) = N_2^2(A(P)) - \sum_{k=1}^{\mu n} |\lambda_k(A(P))|^2 = g_1^2(P)$$

where

$$g_1(P) = [Trace\,(\sum_{k=1}^{\mu} A_k A_k^*) + n(\mu-1) - \sum_{k=1}^{\mu n} |z_k(P)|^2]^{1/2}.$$

Furthermore, simple calculations show that

$$\|A(P) - A(\tilde{P})\|_{\mu n} = q(P, \tilde{P}),$$

where

$$q(P, \tilde{P}) = \left[\sum_{k=1}^{\mu} \|A_k - \tilde{A}_k\|_n^2 \right]^{1/2}.$$

Now Theorem 2.4 implies $sv_{A(P)}(\tilde{A}(P)) \le r(P, \tilde{P})$ where $r(P, \tilde{P})$ is the unique nonnegative root of the algebraic equation

$$y^{\mu n} = q(P, \tilde{P}) \sum_{j=0}^{\mu n-1} \frac{y^{\mu n-j-1} g_1^j(P)}{\sqrt{j!}}. \tag{10.3}$$

Since $\Sigma(P) = \sigma(A(P))$ and $\Sigma(\tilde{P}) = \sigma(A(\tilde{P}))$ we have proved the following result.

Theorem 3.8 *Let P and $\tilde{P}$ be defined by* (10.1). *Then for any characteristic value $z(\tilde{P})$ of $\tilde{P}(z)$, there is a characteristic value $z(P)$ of $P(z)$, such that*

$$|z(P) - z(\tilde{P})| \le r(P, \tilde{P}),$$

where $r(P, \tilde{P})$ is the unique positive root of the algebraic equation (10.3).

Note that

$$Trace\ A^2(P) = Trace\ (A_1^2 - 2A_2).$$

Then

$$\sum_{k=1}^{\mu n} |z_k(P)|^2 \ge |\sum_{k=1}^{\mu n} z_k^2(P)| = |Trace(A(P))^2| = |Trace\ (A_1^2 - 2A_2)|.$$

So

$$g_1(P) \le \left[Trace \left(\sum_{k=1}^{\mu} A_k A_k^* \right) + n(\mu - 1) - |Trace\ (A_1^2 - 2A_2)| \right]^{1/2}.$$

To estimate $r(P, \tilde{P})$ put $y = xg_1(P)$ into (10.1). Then we obtain the equation

$$x^{\mu n} = \frac{q(P, \tilde{P})}{g_1(P)} \sum_{j=0}^{\mu n-1} \frac{x^{\mu n-j-1}}{\sqrt{j!}}.$$

Since

$$\max_{j=0,\dots,\mu n-1} \sqrt[j+1]{\frac{q(P, \tilde{P})}{g_1(P)}} = \frac{q(P, \tilde{P})}{g_1(P)},$$

if $q(P, \tilde{P}) \ge g_1(P)$, and

$$\max_{j=0,\dots,\mu n-1} \sqrt[j+1]{\frac{q(P, \tilde{P})}{g_1(P)}} = \sqrt[\mu n]{\frac{q(P, \tilde{P})}{g(P)}},$$

if $q(P,\tilde{P}) \le g_1(P)$, applying Lemma 2.2 we get the inequality $r(P,\tilde{P}) \le \delta(P,\tilde{P})$, where

$$\delta(P,\tilde{P}) := \begin{cases} 2q(P,\tilde{P}) & \text{if} \quad q(P,\tilde{P}) \ge g_1(P), \\ 2g_1^{1-1/\mu n}(P)[q(P,\tilde{P})]^{1/\mu n} & \text{if} \quad q(P,\tilde{P}) \le g_1(P) \end{cases}.$$

Now Theorem 3.8 ensures the following

Corollary 3.8 *Let P and $\tilde{P}$ be defined by* (10.1)*. Then for any characteristic value $z(\tilde{P})$ of $\tilde{P}(z)$, there is a characteristic value $z(P)$ of $P(z)$, such that*

$$|z(P) - z(\tilde{P})| \le \delta(P,\tilde{P}).$$

Let us consider perturbations of vector difference equations with triangular matrix coefficients. Let

$$\tilde{A}_j = (a_{sk}^{(j)})_{s,k=1}^n \quad (j = 1, \ldots, \mu)$$

be arbitrary $n \times n$-matrices. In addition, let W_j^+, W_j^- and S_j be the upper nilpotent, lower nilpotent, and diagonal parts of $\tilde{A}_j$, respectively. So

$$\tilde{A}_j = W_j^+ + W_j^- + S_j.$$

Take $A_j = W_j^+ + S_j$. That is, A_j is the upper triangular part of $\tilde{A}_j$, and according to (10.1) $P(z)$ is the upper triangular part of $\tilde{P}(z)$. Therefore

$$q(P,\tilde{P}) = q_t := \left[\sum_{k=1}^{\mu} \| W_k^- \|^2 \right]^{1/2}.$$

In addition, since P is triangular, it is not hard to show that

$$g_1(P) = g_t := \sqrt{\sum_{j=1}^{\mu} N_2^2(W_j^+) + (\mu - 1)n}.$$

Under consideration, we have $\delta(P,\tilde{P}) = \delta_t$ where

$$\delta_t := \begin{cases} 2q_t & \text{if} \quad q_t \ge g_t, \\ 2g_t^{1-1/\mu n}[q_t]^{1/\mu n} & \text{if} \quad q_t \le g_t(P) \end{cases}.$$

Because of Corollary 3.8, we get

Corollary 3.9 *All the characteristic values of the matrix polynomial $\tilde{P}(z) = z^\mu I_n + \tilde{A}_1 z^{\mu-1} + \cdots + \tilde{A}_\mu$ with coefficients $\tilde{A}_j = (a_{sk}^{(j)})_{s,k=1}^n$ $(j = 1, \ldots, \mu)$ lie in the* *union of the sets*

$$\{z \in \mathbf{C} : |z - R_{jk}| \le \delta_t\} \ \ (j = 1, \dots, \mu;\ k = 1, \dots, n),$$

where R_{jk} $(j = 1, \dots, \mu)$ *are the roots of the diagonal polynomials*

$$z^\mu + a_{kk}^{(1)} z^{\mu-1} + \cdots + a_{kk}^{(\mu)} \ \ (k = 1, \dots, n).$$

For instance, let us consider the matrix second order polynomial $\tilde{P}(\lambda) = z^2 I_n + \tilde{A}_1 z + \tilde{A}_2$, where

$$\tilde{A}_j = (a_{sk}^{(j)})_{s,k=1}^n \ \ (j = 1, 2).$$

Then

$$R_{1k} = -\frac{a_{kk}^{(1)}}{2} + \sqrt{(a_{kk}^{(1)}/2)^2 - a_{kk}^{(2)}}$$

and

$$R_{2k} = -\frac{a_{kk}^{(1)}}{2} - \sqrt{(a_{kk}^{(1)}/2)^2 - a_{kk}^{(2)}}.$$

Now, we can directly apply Corollary 3.9.

3.11 Perturbations of Characteristic Determinants

Let $A(z)$ and $B(z)$ be entire matrix functions of z (matrices whose entries are entire functions). Since $det\ A(z)$ is a combinations of products and sums of the entries, then $det\ A(z)$ and $det\ B(z)$ are entire functions. $A(z)$ is said to be stable if all its zeros are in $\{z \in \mathbf{C} : Re\ z < 0\}$.

Lemma 3.9 *Assume that* $A(z)$ *is stable and*

$$|det\ A(i\omega) - det\ B(i\omega)| < |det\ A(i\omega)| \ \ (\omega \in \mathbf{R}).$$

Then $B(z)$ *is also stable.*

Proof Since $A(z)$ is stable $det\ A(z)$ does not have zeros on the closed set $\{z \in \mathbf{C} : Re\ z \ge 0\}$. So due to the classical Rouche Theorem, one can assert that $det\ B(z)$ does not have zeros on that set, as claimed. □

Let C and C_1 be $n \times n$-matrices. Then as it is shown in Sect. 2.10,

$$|det\ C - det\ C_1| \le n\|C - C_1\|_n M_2^{n-1}, \tag{11.1}$$

where $M_2 := \max\{\|C\|_n, \|C_1\|_n\}$.

Now let

$$H(\tau) = I - \int_0^\eta \tau^{-s} dR_0(s) \quad \text{and} \quad H_1(\tau) = I - \int_0^\eta \tau^{-s} dR_1(s),$$

where $R_0(s)$ is the same as above and $R_1(s) = (r_{ij}^1(s))_{i,j=1}^n$ is a real $n \times n$-matrix-valued function defined on a finite segment $[0, \eta]$ whose entries have bounded variations and finite numbers of jumps. Recall that $H(z)$ is said to be Schur stable if all its zeros are in the open unit disc $\{z \in \mathbf{C} : |z| < 1\}$.

Clearly, $H(e^z)$ and $H_1(e^z)$ are entire matrix functions.

If $|\tau| = 1$, then

$$\|H(\tau)\|_n \le 1 + \int_0^\eta \|dR_0(s)\|_n = 1 + var(R_0)$$

and

$$\|H_1(\tau)\|_n \le 1 + var(R_1).$$

In addition,

$$\|H(\tau) - H_1(\tau)\|_n \le \int_0^\eta \|d(R_0(s) - R_1(s))\|_n = var\,(R_1 - R_0) \quad (|\tau| = 1).$$

Here $var\,(R_1 - R_0)$ is the spectral norm of the matrix $var\,(R_1 - R_0)$, whose entries are $var\,(r_{ij}^1 - r_{ij})$.

Making use (11.1), we obtain

$$|det\ H(\tau) - det\ H_1(\tau)| \le \Delta \quad (|\tau| = 1), \tag{11.2}$$

where

$$\Delta := n\ var\,(R_1 - R_0)[\max\{var\,(R_0), var\,(R_1)\}]^{n-1}.$$

Now Lemma 3.6 implies

Lemma 3.10 *Let $H(\tau)$ be Schur stable and the condition $|det\ \ H(\tau)| > \Delta\,(|\tau| = 1)$ hold. Then $H_1(\tau)$ is also Schur stable and*

$$|det\ H(\tau)| > |det\ H(\tau)| - \Delta \quad (|\tau| = 1).$$

For instance let

$$H(\tau) = I - A\tau^{-h} \quad \text{and} \quad H_1(\tau) = I - A_1\tau^{-h_1},$$

where A, A_1 are matrices, h, h_1 are positive constants. Then

$$var\ (R_0) = \|A\|_n, var\ (R_1) = \|A_1\|_n, var\ (R_1 - R_0) = \|A - A_1\|_n.$$

In addition, $\lambda_k(H(\tau)) = 1 - \lambda_k(A)\tau^{-h}$. The characteristic values are $z_k(H) = \lambda_k^{1/h}(A)$. So if A is Schur stable, then $H(\tau)$ is also Schur stable and

$$det\ H(\tau) = \prod_{k=1}^{n}(1 - \lambda_k(A)\tau^{-h}).$$

Hence,

$$|det\ H(\tau)| \geq \prod_{k=1}^{n}(1 - |\lambda_k(A)|)\ \ (|\tau| = 1).$$

In addition,

$$\Delta = n\|A - A_1\|_n[\max\{\|A\|_n, \|A_1\|_n\}]^{n-1}$$

So, if A is Schur stable and

$$\prod_{k=1}^{n}(1 - |\lambda_k(A)|) > n\|A - A_1\|_n[\max\{\|A\|_n, \|A_1\|_n\}]^{n-1},$$

then $H_1(\tau)$ is also Schur stable.

3.12 Comments

The material of the present chapter is probably new.

The importance of continuous time difference and difference-delay equations in applications is very well explained in several books (Hale and Lunel 1993; Kolmanovskii and Myshkis 1999; Niculescu 2001). These equations appear in economics (Cheng and Lin 2005), gas dynamics, propagation models (Niculescu 2001). They also appear as the internal dynamics of non-linear differential delay systems, when these systems are input-output linearized and controlled, cf. Pepe (2003). For us it is important that the difference-delay operators with continuous time form the principal parts of neutral type functional differential equations.,

Results concerning the stability of continuous time difference-delay equations in the linear case, can be found, in particular, in Hale and Lunel (1993) (see Chap. 9 of that book on the equations of neutral type). In particular, in Sect. 9.6 of Hale and Lunel (1993), the stability of linear, time invariant, continuous time difference-delay equations is studied, taking into account small variations that may occur in the delays.

In the book (Sharkovsky et al. 1993), asymptotic properties of the solutions of scalar difference equations with continuous argument are treated.

In the paper (Korenevskii and Kaizer 1998) sufficient algebraic coefficient conditions are established for the asymptotic stability of solutions of systems of linear difference equations with continuous time and delay in the case of a rational correlation between delays. The authors formulate the conditions of asymptotic stability in terms of matrix algebraic Lyapunov equations.

In the paper (Gil' and Cheng 2007) the authors consider the following difference delay equation with continuous time in a Euclidean space $\mathbf{C}^n$:

$$x(t) = \sum_{k=1}^{m} A_k x(t - h_k) + F(t, x(t - h_1, \ldots, x(t - h_m)) + f(t), \ t > 0, \quad (12.1)$$

where f is a given vector valued function, a mapping $F : [0, \infty) \times \mathbf{C}^n \to \mathbf{C}^n$ satisfies certain conditions pointed below, A_k are n by n matrices and $h_k, \ k = 1, \ldots, m$ are nonnegative numbers. Solution estimates to Eq. (12.1) are derived. These estimates provide conditions for the stability and boundedness of solutions.

In the interesting paper (Pepe and Verriest 2003), a Lyapunov-type stability analysis is given for coupled delay differential and continuous time difference equations is given. The main contribution of the paper (Pepe and Verriest 2003) is a two-step algorithm for the stability analysis of such systems. At the first step, only stability is checked, while at the second step, the Lyapunov stability of the difference part of the system is analyzed. In these two steps, the stability of the whole system can be checked.

To the best of our knowledge, however, stability of general continuous time difference equations has not been extensively investigated in the available literature.

Chapter 4
Linear Differential Delay Equations

This chapter is devoted to vector linear differential-delay equations (DDEs). We derive estimates for the L^p- and C-norms of the Cauchy operators of autonomous and time-variant differential-delay equations. These estimates in the sequel enable us to establish stability conditions for linear and nonlinear neutral type functional differential equations.

In this chapter we again write $L^p(a, b)$ instead of $L^p([a, b]; \mathbf{C}^n)$ and $C(a, b)$ instead of $C([a, b]; \mathbf{C}^n)$.

4.1 Homogeneous Autonomous Equations

Again $R_0(\tau) = (r_{jk}(\tau))_{j,k=1}^n$ is a real $n \times n$-matrix-valued function defined on a finite segment $[0, \eta]$, having bounded variation and a finite number of jumps.

Consider the problem

$$\dot{y}(t) = \int_0^\eta dR_0(\tau)y(t-\tau) \ (t > 0), \tag{1.1}$$

$$y(t) = \phi(t) \text{ for } -\eta \le t \le 0, \tag{1.2}$$

where $\phi(t)$ is a given continuous vector-valued function defined on $[-\eta, 0]$. Besides, $\dot{y}(t) = dy(t)/dt$, $t > 0$, and $\dot{y}(0)$ means the right derivative of y at zero.

The following equation is an example of Eq. (1.1):

$$\dot{y}(t) = \int_0^\eta A(s)y(t-s)ds + \sum_{k=1}^m A_k y(t-h_k) \ \ (t > 0; \ m < \infty), \tag{1.3}$$

M.I. Gil', *Stability of Neutral Functional Differential Equations*,
Atlantis Studies in Differential Equations 3, DOI 10.2991/978-94-6239-091-1_4

where

$$0 \le h_1 < h_2 < \cdots < h_m < \eta$$

are constants, $A_k = (a_{ij}^{(k)})_{i,j=1}^n$ are constant matrices and $A(s) = (a_{ij}(s))_{i,j=1}^n$ is an integrable on $[0, \eta]$ matrix function.

A solution $y(t)$ of problem (1.1) and (1.2) is a continuous function $y := [-\eta, \infty) \to \mathbf{C}^n$, such that

$$y(t) = \phi(0) + \int_0^t \int_0^\eta dR_0(\tau) y(s-\tau) ds, \quad t \ge 0, \tag{1.4}$$

and (1.2) holds.

Recall that $\|A\|_n$ is the spectral (operator) norm of a matrix A, $\|v\|_n$ is the Euclidean norm of $v \in \mathbf{C}^n$ and $var\ (R_0)$ is the spectral norm of matrix $\left(var(r_{ij})\right)_{i,j=1}^n$: $var\ (R_0) = \|(var(r_{ij})\|_n$. For Eq. (1.3), each entry of R_0, satisfies the inequality

$$var\ (r_{ij}) \le \int_0^\eta |a_{ij}(s)| ds + \sum_{k=1}^m |a_{ij}^{(k)}|. \tag{1.5}$$

Furthermore, Lemma 1.6 implies

$$\sup_{0 \le s \le T} \| \int_0^\eta dR_0(\tau) u(s-\tau) \|_n \le \sqrt{n}\ var\ (R_0) \sup_{-\eta \le s \le T} \|u(s)\|_n \quad (u \in C(0,T)). \tag{1.6}$$

Let $y(t)$ be a solution to problem (1.1) and (1.2). Put

$$y_C(t) := \sup_{0 \le s \le t} \|y(s)\|_n$$

and take into account that

$$\begin{aligned} \| \int_0^t \int_0^\eta dR_0(\tau) y(s-\tau) ds \|_n &= \| \int_0^\eta \int_{-\tau}^{t-\tau} dR_0(\tau) y(s_1) ds_1 \|_n \\ &\le \| \int_0^\eta \int_{-\tau}^{t-\tau} \|dR_0(\tau)\|_n \|y(s_1) ds_1 \|_n \\ &\le \sqrt{n}\ var\ (R_0) \int_{-\eta}^t \|y(s_1)\| ds_1 \end{aligned}$$

$$= \sqrt{n}\, var\ (R_0)(\int_{-\eta}^{0} \|\phi(s_1)\|ds_1 + \int_0^t \|y(s_1)\|ds_1).$$

Then from (1.4) we deduce that

$$y_C(t) \le const\ \|\phi\|_{C(-\eta,0)} + \sqrt{n}\, var\ (R_0) \int_0^t y_C(s)ds.$$

By the Gronwall lemma we arrive at the inequality

$$\|y(t)\|_n \le y_C(t) \le const\ \|\phi\|_{C(-\eta,0)} e^{t\sqrt{n}\, var\ (R_0)}\ (t \ge 0). \tag{1.7}$$

This estimate provides the existence and uniqueness of solutions. For more details see, for instance, Kolmanovskii and Myshkis (1999).

4.2 Non-homogeneous Autonomous Equations

Consider also the non-homogeneous problem

$$\dot{x}(t) = \int_0^{\eta} dR_0(\tau)x(t-\tau) + f(t)\ \ (t > 0), \tag{2.1}$$

with a locally integrable f and the zero initial condition

$$x(t) = 0 \text{ for } -\eta \le t \le 0. \tag{2.2}$$

A solution of problem (2.1) and (2.2) is a continuous vector function $x(t)$ defined on $[0, \infty)$ and satisfying the equation

$$x(t) = \int_0^t \int_0^{\eta} dR_0(\tau)x(s-\tau)ds + \int_0^t f(s)ds\ (t \ge 0) \tag{2.3}$$

and condition (2.2). Hence,

$$\|x\|_{C(0,t)} \le \int_0^t \|f(s)\|_n ds + \sqrt{n}\, var\ (R_0) \int_0^t \|x\|_{C(0,s)} ds.$$

Now the Gronwall lemma implies the inequality

$$\|x(t)\|_n \leq \|x\|_{C(0,t)} \leq \int_0^t \|f(s)\|_n ds\ exp\ [t\sqrt{n}\ var\ (R_0)].$$

Assume also that f is of exponential growth. Taking into account (1.7), one can apply the Laplace transforms

$$\tilde{y}(z) = \int_0^\infty e^{-zt} y(t)dt \text{ and } \tilde{x}(z) = \int_0^\infty e^{-zt} x(t)dt$$

to solutions of problems (1.1), (1.2), (2.1) and (2.2), respectively, and the integrals converge absolutely in the corresponding half-planes. In addition, inequality (1.7) together with Eq. (1.1) show that also $\dot{y}(t)$ has a Laplace transform at least in $Re\ z > \sqrt{n}\ var\ (R_0)$ given by $z\tilde{y}(z) - \phi(0)$. Taking the Laplace transforms of the both sides of Eq. (1.1), we get

$$\begin{aligned} z\tilde{y}(z) - \phi(0) &= \int_0^\infty e^{-zt} \int_0^\eta dR_0(\tau) y(t-\tau)dt \\ &= \int_0^\eta e^{-\tau z} dR_0(\tau) \left[\int_{-\tau}^0 e^{-zt} y(t)dt + \tilde{y}(z) \right]. \end{aligned}$$

Interchanging the Stieltjes integration with the improper Riemann integration is justified by the Fubini theorem; we thus get

$$K(z)\tilde{y}(z) = \phi(0) + \int_0^\eta e^{-\tau z} dR_0(\tau) \int_{-\tau}^0 e^{-zt} \phi(t)dt, \tag{2.4}$$

where

$$K(z) = zI - \int_0^\eta e^{-\tau z} dR_0(\tau). \tag{2.5}$$

The matrix-valued function $K(z)$ is called *the characteristic matrix-valued function* to Eq. (1.1) and $det\ K(z)$ is called *the characteristic determinant* of Eq. (1.1). A zero of the characteristic determinant $det\ K(z)$ is called *a characteristic value of* $K(.)$ and $\lambda \in \mathbf{C}$ is *a regular value* of $K(.)$ if $det\ K(\lambda) \neq 0$.

In the sequel it is assumed that all the characteristic values of $K(.)$ *are in the open left half-plane* C_-. That is, Eq. (1.1) is exponentially stable (Kolmanovski and Myshkis 1999).

Below we also suggest some conditions that provide this assumption.

Applying the inverse Laplace transform to (2.4), we obtain

$$y(t) = \frac{1}{2\pi} \int_{-\infty}^{\infty} e^{i\omega t} K^{-1}(i\omega) \left[\phi(0) + \int_0^{\eta} dR_0(\tau) \int_{-\tau}^{0} e^{-i\omega(s+\tau)} \phi(s) ds \right] d\omega \tag{2.6}$$

for $t \geq 0$.

Furthermore, apply the Laplace transform to problem (2.1) and (2.2). Then we easily get

$$\tilde{x}(z) = K^{-1}(z) \tilde{f}(z) \tag{2.7}$$

for all regular z. Here $\tilde{f}(z)$ is the Laplace transforms to f. Applying the inverse Laplace transform to (2.7), we get the following equality:

$$x(t) = \int_0^t G(t-s) f(s) ds \quad (t \geq 0), \tag{2.8}$$

where

$$G(t) = \frac{1}{2\pi} \int_{-\infty}^{\infty} e^{i\omega t} K^{-1}(i\omega) d\omega. \tag{2.9}$$

Clearly, the matrix-valued function $G(t)$ satisfies (1.1). Moreover,

$$G(0+) = I, \;\; G(t) = 0 \; (t < 0). \tag{2.10}$$

$G(t)$ is called *the fundamental solution* of Eq. (1.1). Formula (2.8) is *the Variation of Constants formula* to problem (2.1) and (2.2).

Note that for Eq. (1.3) we have

$$K(z) = zI - \int_0^{\eta} e^{-sz} A(s) ds - \sum_{k=1}^{m} e^{-h_k z} A_k. \tag{2.11}$$

The following result can be directly deduced from (2.6): let $y(t)$ be a solution of the time-invariant homogeneous problem (1.1) and (1.2). Then

$$y(t) = G(t)\phi(0) + \int_0^\eta \int_{-\tau}^0 G(t-\tau-s)dR_0(\tau)\phi(s)ds \quad (t \geq 0). \tag{2.12}$$

4.3 Estimates for Characteristic Matrices

Let A be an $n \times n$-matrix. Recall that the quantity

$$g(A) = [N_2^2(A) - \sum_{k=1}^n |\lambda_k(A)|^2 \,]^{1/2}$$

is introduced in Sect. 2.1. Here $\lambda_k(A)$, $k = 1, \ldots, n$, are the eigenvalues of A, counted with their multiplicities; $N_2(A) = (Trace\ AA^*)^{1/2}$ is the Frobenius (Hilbert-Schmidt norm) of A, A^* is adjoint to A. As it is shown in Sect. 2.1, the following relations are valid:

$$g^2(A) \leq N_2^2(A) - |Trace\ A^2| \text{ and } g^2(A) \leq \frac{N_2^2(A - A^*)}{2} = 2N_2^2(A_I), \tag{3.1}$$

where $A_I = (A - A^*)/2i$. Moreover,

$$g(e^{it}A + zI) = g(A) \quad (t \in \mathbf{R};\, z \in \mathbf{C}). \tag{3.2}$$

If A is a normal matrix: $AA^* = A^*A$, then $g(A) = 0$. Put

$$B(z) = \int_0^\eta e^{-z\tau} dR_0(\tau) \quad (z \in \mathbf{C}).$$

In particular, for Eq. (1.3) we have

$$B(z) = \int_0^\eta e^{-sz} A(s)ds + \sum_{k=1}^m e^{-h_k z} A_k \tag{3.3}$$

and

$$g(B(i\omega)) \leq N_2(B(i\omega)) \leq \int_0^\eta N_2(A(s))ds + \sum_{k=1}^m N_2(A_k) \quad (\omega \in \mathbf{R}). \tag{3.4}$$

Below, under various assumptions, we suggest the sharper estimates for $g(B(i\omega))$.

According to Corollary 2.1, the inequality

$$\|A^{-1}\|_n \leq \sum_{k=0}^{n-1} \frac{g^k(A)}{\sqrt{k!}r_l^{k+1}(A)}$$

is valid for any invertible matrix A, where $r_l(A)$ is the smallest absolute value of the eigenvalues (the lower spectral radius) of A.

Hence we arrive at the inequality

$$\|[K(z)]^{-1}\|_n \leq \Gamma(K(z)) \ (z \in \mathbf{C}), \tag{3.5}$$

where

$$\Gamma(K(z)) = \sum_{k=0}^{n-1} \frac{g^k(B(z))}{\sqrt{k!}d^{k+1}(K(z))}$$

and $d(K(z))$ is the smallest absolute value of the eigenvalues of matrix $K(z)$ for a fixed z:

$$d(K(z)) = \min_{k=1,\dots,n} |\lambda_k(K(z))|.$$

If $B(z)$ is a normal matrix, then $g(B(z)) = 0$, and

$$\|[K(z)]^{-1}\|_n \leq \frac{1}{d(K(z))}.$$

For example that inequality holds, if $K(z) = zI - A_0 e^{-z\eta}$, where A_0 is a Hermitian matrix.

Denote

$$\theta(K) := \sup_{-2\,var(R_0) \leq \omega \leq 2\,var(R_0)} \|K^{-1}(i\omega)\|_n.$$

Lemma 4.1 *The equality*

$$\sup_{-\infty \leq \omega \leq \infty} \|K^{-1}(i\omega)\|_n = \theta(K)$$

is valid.

Proof We have

$$K(0) = \int_0^\eta dR_0(s) = \int_0^\eta dR_0(s) = R_0(\eta) - R_0(0).$$

So

$$\|K(0)\|_n = \|R_0(\eta) - R_0(0)\|_n \le var(R_0),$$

and therefore,

$$\|K^{-1}(0)\|_n \ge \frac{1}{var(R_0)}.$$

Simple calculations show that

$$\|\int_0^\eta e^{-i\omega\tau} dR_0(\tau)\|_n = \|(\int_0^\eta e^{-i\omega\tau} dr_{jk}(\tau))_{j,k=1}^n\|_n$$
$$\le \|(\int_0^\eta |dr_{jk}(\tau)|)_{j,k=1}^n\|_n = var(R_0) \quad (\omega \in \mathbf{R})$$

and

$$\|K(i\omega)v\|_n \ge (|\omega| - var\ (R_0))\|v\|_n \ge var(R_0)\|v\|_n$$
$$(\omega \in \mathbf{R};\ |\omega| \ge 2var(R_0);\ v \in \mathbf{C}^n).$$

So

$$\|K^{-1}(i\omega)\|_n \le \frac{1}{var(R_0)} \le \|K^{-1}(0)\|_n \quad (|\omega| \ge 2var(R_0)).$$

Thus the maximum of $\|K^{-1}(i\omega)\|_n$ is attained inside the segment $[-2\ var(R_0), 2\ var(R_0)]$, as claimed. □

By (3.5) and the previous lemma we have the inequality $\theta(K) \le \hat{\theta}(K)$, where

$$\hat{\theta}(K) := \sup_{|\omega| \le 2var(R_0)} \Gamma(K(i\omega)).$$

Denote

$$g_B := \sup_{\omega \in [-2var(R_0), 2var(R_0)]} g(B(i\omega))$$

and

$$d_K := \inf_{\omega \in [-2var(R_0), 2var(R_0)]} d(K(i\omega)).$$

Then we obtain our next result.

Corollary 4.1 *The inequalities*

$$\theta(K) \le \hat{\theta}(K) \le \Gamma_0(K),$$

are true, where

$$\Gamma_0(K) := \sum_{k=0}^{n-1} \frac{g_B^k}{\sqrt{k!}d_K^{k+1}}.$$

4.4 The Cauchy Operator of an Autonomous Equation

4.4.1 L^2-norm Estimates for the Cauchy Operator

Recall that all the characteristic values of $K(.)$ are assumed to be in the open left half-plane C_-.

Lemma 4.2 *Let $x(t)$ be a solution of problem* (2.1) *and* (2.2) *with $f \in L^2(0, \infty)$. Then*

$$\|x\|_{L^2(0,\infty)} \le \theta(K)\|f\|_{L^2(0,\infty)}.$$

Proof Thanks to (2.7),

$$\|\hat{x}^{-1}(i\omega)\|_{L^2(-\infty,\infty)} \le \sup_{-\infty \le \omega \le \infty} \|K^{-1}(i\omega)\|_n \|\hat{f}^{-1}(i\omega)\|_{L^2(-\infty,\infty)}$$

Now Lemma 4.1 implies

$$\|\hat{x}^{-1}(i\omega)\|_{L^2(-\infty,\infty)} \le \theta(K)\|\hat{f}^{-1}(i\omega)\|_{L^2(-\infty,\infty)}.$$

But due to the Parseval equality, we can write

$$\|f\|^2_{L^2(0,\infty)} = \frac{1}{2\pi}\|\hat{f}(i\omega)\|^2_{L^2(-\infty,\infty)} \text{ and } \|x\|^2_{L^2(0,\infty)} = \frac{1}{2\pi}\int_{-\infty}^{\infty} \|\hat{x}(i\omega)\|_n^2 d\omega.$$

This proves the required result. □

The the operator $\hat{G}$ defined by

$$\hat{G}f(t) = \int_0^t G(t-s)f(s)ds,$$

where $G(t)$ is the fundamental solution to Eq. (2.1) is called *the Cauchy (resolvent) operator* to Eqs. (1.1) and (2.1). We thus arrive at

Theorem 4.1 *The Cauchy operator to the differential delay Eq.* (1.1) *satisfies the inequality*

$$\|\hat{G}\|_{L^2(0,\infty)} \leq \theta(K).$$

Now Corollary 4.1 implies

Corollary 4.2 *One has*

$$\|\hat{G}\|_{L^2(0,\infty)} \leq \hat{\theta}(K) \leq \Gamma_0(K).$$

4.4.2 Integrals of Characteristic Functions

Note that, making use the Parseval equality for matrix valued functions (see Sect. 1.10), it is not hard to check that

$$\int_0^\infty N_2^2(G(t))dt = \frac{1}{2\pi}\int_{-\infty}^\infty N_2^2(K^{-1}(i\omega))d\omega. \tag{4.1}$$

However calculations of the integral

$$\int_{-\infty}^\infty N_2^2(K^{-1}(i\omega))d\omega$$

is often a difficult task. Because of this, in the next lemma we suggest an estimate for that integral.

Lemma 4.3 *The inequality*

$$\|K^{-1}(i\omega)\|_{L^2(-\infty,\infty)} := \left(\int_{-\infty}^\infty \|K^{-1}(i\omega)\|_n^2 d\omega\right)^{1/2} \leq \sqrt{2\theta^2(K)var(R_0) + 4\theta(K)}$$

is valid.

Proof We have

$$\|K(i\omega)u\|_n = \|i\omega I u - \int_0^\eta e^{-i\omega\tau} dR_0(\tau)u\|_n$$

$$\geq (|\omega| - \int_0^\eta \|dR_0(\tau)\|_n)\|u\|_n$$
$$= (|\omega| - var\ (R_0))\|u\|_n\ (\omega \in \mathbf{R};\ |\omega| \geq var(R_0);\ \ u \in \mathbf{C}^n).$$

Therefore,

$$\|K^{-1}(i\omega)\|_n \leq \frac{1}{|\omega| - var(R_0)}\ \ (|\omega| > var(R_0)). \tag{4.2}$$

For the brevity, in this proof put

$$m(\omega) = \|K^{-1}(i\omega)\|_n,\ v = var(R_0),\ \theta = \theta(K) \text{ and } a = v + \frac{1}{\theta}\ \ (\omega \in \mathbf{R}).$$

So

$$m(\omega) \leq \frac{1}{|\omega| - v}\ (|\omega| > v).$$

We should estimate the integral

$$J = \int_{-\infty}^{\infty} m^2(s)ds = J_1 + J_2,$$

where

$$J_1 = \int_{-a}^{a} m^2(s)ds,\ J_2 = \int_{-\infty}^{-a} m^2(s)ds + \int_a^{\infty} m^2(s)ds.$$

Due to Lemma 4.1

$$\|m(\omega)\|_n \leq \theta \text{ for all } \omega \in \mathbf{R}. \tag{4.3}$$

Thus

$$J_1 \leq 2a\theta^2 = 2\theta^2(v + 1/\theta)$$

and

$$J_2 \leq 2\int_a^{\infty} (s - v)^{-2}ds = 2(a - v)^{-1} = 2\theta.$$

Thus $J \leq 2(\theta^2 v + 2\theta)$, as claimed. □

Since,

$$\|K(0)\|_n = \|\int_0^\eta dR_0(s)\|_n \le var(R_0)$$

we have

$$\theta(K) \ge \|K^{-1}(0)\|_n \ge \frac{1}{var(R_0)}.$$

Or

$$\theta(K)var(R_0) \ge 1. \tag{4.4}$$

Now the previous lemma implies

$$\|K^{-1}(i\omega)\|_{L^2(-\infty,\infty)} \le \theta(K)\sqrt{6\,var(R_0)}. \tag{4.5}$$

Furthermore, let $vd(R_0)$ *mean the spectral norm of the matrix* $(vd(r_{jk}))_{j,k=1}^n$, *whose entries are*

$$vd(r_{ij}) = \int_0^\eta s|dr_{jk}(s)|$$

(see also Sect. 1.8). For example, in the case of the equation

$$\dot{y}(t) = A_1 y(t-h_1) + A_1 y(t-h_2)$$

with constant matrices A_1 and A_2 we have

$$vd(R_0) \le \|A_1\|_n h_1 + \|A_2\|_n h_2.$$

In the case of the equation

$$\dot{y}(t) = \int_0^1 A(s)y(t-s)ds$$

with an integrable matrix $A(t)$ we have

$$vd(R_0) \le \int_0^1 s\|A(s)\|_n ds.$$

We need the following lemma.

Lemma 4.4 *The inequality*

$$\left\|\frac{dK^{-1}(i\omega)}{d\omega}\right\|_{L^2(-\infty,\infty)} := \left(\int_{-\infty}^{\infty}\left\|\frac{dK^{-1}(i\omega)}{d\omega}\right\|_n^2 d\omega\right)^{1/2}$$
$$\leq (1+vd(R_0))\sqrt{2(\theta^4(K)var(R_0)+\frac{4}{3}\theta^3(K))}$$

is true.

Proof We have

$$\frac{dK^{-1}(z)}{dz} = -K^{-1}(z)\frac{dK(z)}{dz}K^{-1}(z).$$

In addition,

$$\frac{dK(i\omega)}{d\omega} = iI + i\int_0^{\eta}\tau e^{-i\tau\omega}dR_0(\tau)$$

and therefore, due to Lemma 1.7,

$$\left\|\frac{dK(i\omega)}{d\omega}\right\|_n \leq 1+vd(R_0).$$

So

$$\left\|\frac{dK^{-1}(i\omega)}{d\omega}\right\|_n \leq \|K^{-1}(i\omega)\|_n^2(1+vd(R_0)),$$

and consequently,

$$\left\|\frac{dK^{-1}(i\omega)}{d\omega}\right\|_{L^2(-\infty,\infty)}^2 \leq (1+vd(R_0))^2\int_{-\infty}^{\infty}\|K^{-1}(i\omega)\|_n^4 d\omega \tag{4.6}$$

Again use the notations

$$m(\omega) = \|K^{-1}(i\omega)\|_n,\ v = var(R_0),\ \theta = \theta(K) \text{ and } a = v + \frac{1}{\theta}\quad (\omega \in \mathbf{R}).$$

According to (4.6) we need to estimate the integral

$$j(K) = \int_{-\infty}^{\infty} m^4(s)ds = j_1 + j_2$$

where

$$j_1 = \int_{-a}^{a} m^4(s)ds, \; j_2 = \int_{-\infty}^{-a} m^4(s)ds + \int_{a}^{\infty} m^4(s)ds.$$

Due to (4.2) and (4.3)

$$j_1 \le 2\theta^4 a = 2\theta^4(v + \frac{1}{\theta})$$

and

$$j_2 \le 2\int_{a}^{\infty} (s-v)^{-4} ds = \frac{2}{3(a-v)^3} = \frac{2}{3}\theta^3.$$

Thus

$$j(K) \le 2(\theta^4 v + \frac{4}{3}\theta^3).$$

According to (4.6) this proves the lemma. □

Inequality (4.4) and the later lemma imply

$$\left\| \frac{dK^{-1}(i\omega)}{d\omega} \right\|_{L^2(-\infty,\infty)} \le \theta^2(K)(1 + vd(R_0))\sqrt{\frac{14}{3} var(R_0)}. \tag{4.7}$$

4.4.3 Integrals of Fundamental Solutions

Due to (4.1) and the Parseval equality we obtain

$$\int_0^\infty N_2^2(G(t))dt = \frac{1}{2\pi}\int_{-\infty}^{\infty} N_2^2(K^{-1}(i\omega))d\omega \le \frac{n}{2\pi}\int_{-\infty}^{\infty} \|K^{-1}(i\omega)\|_n^2 d\omega.$$

Moreover, by the Parseval equality

$$\int_0^\infty \|G(t)u\|_n^2 dt = \frac{1}{2\pi}\int_{-\infty}^{\infty} \|K^{-1}(i\omega)u\|_n^2 d\omega \quad (u \in \mathbf{C}^n).$$

Put

$$W(K) = \sqrt{\frac{n}{\pi}(\theta^2(K)var(R_0) + 2\theta(K))}.$$

Now Lemma 4.3 implies

Corollary 4.3 *One has*

$$\int_0^\infty N_2^2(G(t))dt \le W^2(K)$$

and

$$\int_0^\infty \|G(t)u\|_n^2 dt \le \frac{1}{n}W^2(K) \quad (u \in \mathbf{C}^n;\ \|u\|_n = 1).$$

According to (4.4) the latter corollary implies

$$\int_0^\infty N_2^2(G(t))dt \le W^2(K) \le \theta^2(K)\frac{3n\, var(R_0)}{\pi} \le \theta^2(K)n\, var(R_0). \tag{4.8}$$

Furthermore, the Laplace transform to $tG(t)$ is $-dK^{-1}(z)/dz$. So by the Parseval equality we get

$$\int_0^\infty t^2 N_2^2(G(t))dt = \frac{1}{2\pi}\int_{-\infty}^\infty N_2^2(dK^{-1}(is)/ds)ds \le n\frac{1}{2\pi}\left\|\frac{dK^{-1}(i\omega)}{d\omega}\right\|^2_{L^2(-\infty,\infty)}.$$

Now Lemma 4.4 implies

$$\int_0^\infty t^2\|G(t)\|_n^2 dt \le \frac{n}{\pi}(1 + vd(R_0))^2(\theta^4(K)var(R_0) + \frac{4}{3}\theta^3(K)).$$

Hence, according to (4.4)

$$\int_0^\infty t^2\|G(t)\|_n^2 dt \le \frac{7n}{3\pi}(1 + vd(R_0))^2\theta^4(K)var(R_0). \tag{4.9}$$

4.4.4 An Estimate for the C-norm of the Fundamental Solution

We need the inequality

$$\| \int_0^\eta dR_0(s)f(t-s)\|_{L^2(0,\infty)} \le var(R_0)\|f\|_{L^2(-\eta,\infty)} \quad (f \in L^2(-\eta,\infty)) \tag{4.10}$$

proved in Sect. 1.8. Take into account that the fundamental solution satisfies the Eq. (1.1). So

$$\dot{G}(t) = \int_0^\eta dR_0(\tau)G(t-\tau).$$

Hence,

$$\|\dot{G}\|_{L^2(0,\infty)} \le \|G\|_{L^2(0,\infty)}\ var\ (R_0).$$

Now Corollary 4.3 implies

$$\|\dot{G}\|_{L^2(0,\infty)} \le W(K)\ var\ (R_0). \tag{4.11}$$

We need also the following simple result.

Lemma 4.5 *Let* $f \in L^2(0,\infty)$ *and* $\dot{f} \in L^2(0,\infty)$. *Then*

$$\|f\|^2_{C(0,\infty)} \le 2\|f\|_{L^2(0,\infty)}\|\dot{f}\|_{L^2(0,\infty)}.$$

Proof Obviously,

$$\|f(t)\|_n^2 = -\int_t^\infty \frac{d\|f(s)\|_n^2}{ds}ds = -2\int_t^\infty \|f(s)\|_n \frac{d\|f(s)\|_n}{ds}ds.$$

Taking into account that

$$|\frac{d}{ds}\|f(s)\|_n| \le \|\dot{f}(s)\|_n,$$

we get the required result due to Schwarz's inequality. □

By the previous lemma and (4.11) we get the following inequality:

$$\|Gu\|^2_{C(0,\infty)} \le 2\|Gu\|^2_{L^2(0,\infty)}\ var\ (R_0) \quad (u \in \mathbf{C}^n).$$

Take t_0 such that

$$\|G\|_{C(0,\infty)} = \|G(t_0)\|_n.$$

and $u_0 \in \mathbf{C}^n$ with $\|u_0\|_n = 1$ such that $\|G(t_0)u_0\|_n = \|G(t_0)\|_n$. Then we have

$$\|G\|^2_{C(0,\infty)} \leq 2\|Gu_0\|^2_{L^2(0,\infty)} \; var\ (R_0).$$

Now by Corollary 4.3 at once we obtain the following result.

Lemma 4.6 *The inequality* $\|G\|_{C(0,\infty)} \leq a_0(K)$, *is valid, where*

$$a_0(K) := \sqrt{2var\ (R_0)}W(K).$$

Making use inequality (4.8) we get

$$a_0(K) \leq \theta(K)\ var(R_0)\sqrt{\frac{6n}{\pi}} \leq \sqrt{2n}\theta(K)\ var(R_0).$$

4.4.5 C- and L^p-norms of the Cauchy Operator

Recall that all the characteristic values of K are assumed to be in C_-.

Lemma 4.7 *The fundamental solution G to Eq.* (1.1), *satisfies the inequality*

$$\|G\|_{L^1(0,\infty)} \leq \sqrt{\pi\|G\|_{L^2(0,\infty)}\|Y\|_{L^2(0,\infty)}} \tag{4.12}$$

where $Y(t) = tG(t)$.

Proof This result is due to the Karlson inequality

$$(\int_0^\infty |f(t)|dt)^4 \leq \pi^2 \int_0^\infty f^2(t)dt \int_0^\infty t^2 f^2(t)dt$$

for a real scalar-valued $f \in L^2[0,\infty)$ with the property $tf(t) \in L^2[0,\infty)$, cf. (Mitrinović et al. 1991, Chap. VIII). □

Inequalities (4.8), (4.9) and the latter lemma imply

Corollary 4.4 *Let the fundamental solution to Eq.* (1.1), *is subject to the estimate* $\|G\|_{L^1(0,\infty)} \leq \zeta(K)$, *where*

$$\zeta(K) := \theta^{3/2}(K)\sqrt{3n\ var(R_0)\ (1 + vd(R_0))}.$$

Recall that some estimates for $\theta(K)$ are given in Sect. 4.3. Additional estimates are suggested below.

Taking into account Theorem 1.5 on the properties of convolutions, we arrive at the main result of the present section.

Theorem 4.2 *The Cauchy operator to the differential delay Eq.* (1.1) *satisfies the inequalities* $\|\hat{G}\|_{L^p(0,\infty)} \leq \zeta(K)$ $(p \geq 1)$ *and* $\|\hat{G}\|_{C(0,\infty)} \leq \zeta(K)$.

Recall that in the case $p = 2$ we have the sharper inequality $\|\hat{G}\|_{L^2(0,\infty)} \leq \theta(K)$.

4.5 Systems with Several Distributed Delays

The aim of this section is to evaluate $g(B(z))$ for the system

$$\dot{y}(t) = \sum_{k=1}^{m} A_k \int_0^{\eta} y(t-s)d\mu_k(s), \tag{5.1}$$

where A_k are constant matrices and $\mu_k(s)$ are scalar nondecreasing functions defined on $[0, \eta]$.

In this case the characteristic matrix function is

$$K(z) = zI - \sum_{k=1}^{m} A_k \int_0^{\eta} e^{-zs} d\mu_k(s).$$

Simple calculations show that

$$var(R_0) \leq \sum_{k=1}^{m} \|A_k\|_n var\ (\mu_k),\ vd(R_0) \leq \sum_{k=1}^{m} \|A_k\|_n \int_0^{\eta} s d\mu_k(s)$$

and

$$g(K(i\omega)) = g(B(i\omega)) \leq \sum_{k=1}^{m} N_2(A_k) var(\mu_k) \quad (\omega \in \mathbf{R}).$$

For instance, the equation

$$\dot{y}(t) = \sum_{k=1}^{m} A_k y(t-h_k) \tag{5.2}$$

can take the form (5.1); besides,

$$K(z) = zI - \sum_{k=1}^{m} A_k e^{-h_k z},\ var(R_0) \le \sum_{k=1}^{m} \|A_k\|_n,\ vd(R_0) \le \sum_{k=1}^{m} mh_k \|A_k\|_n$$

and

$$g(B(i\omega)) \le \sum_{k=1}^{m} N_2(A_k) \quad (\omega \in \mathbf{R}).$$

Under additional conditions, the latter estimate can be improved. For example, if

$$K(z) = zI - A_1 e^{-h_k z} - A_2 e^{-h_2 z}, \tag{5.3}$$

then due to (3.1) and (3.2), for all $\omega \in \mathbf{R}$, we obtain

$$\begin{aligned} g(B(i\omega)) &= g(e^{i\omega h_1} B(i\omega)) = g(A_1 + A_2 e^{(h_1 - h_2)i\omega}) \\ &\le \frac{1}{\sqrt{2}} N_2(A_1 - A_1^* + A_2 e^{(h_1-h_2)i\omega} - A_2^* e^{-(h_1-h_2)i\omega}) \end{aligned}$$

and, consequently,

$$g(B((i\omega)) \le \frac{1}{\sqrt{2}} N_2(A_1 - A_1^*) + \sqrt{2} N_2(A_2). \tag{5.4}$$

Similarly, we get

$$g(B((i\omega)) \le \sqrt{2} N_2(A_1) + \frac{1}{\sqrt{2}} N_2(A_2 - A_2^*) \quad (\omega \in \mathbf{R}). \tag{5.5}$$

4.6 Scalar Autonomous Differential Delay Equations

4.6.1 The General Case

In this section we are going to estimate the characteristic functions and fundamental solutions of some scalar equations. These estimates give us lower bounds for the quantity d_K. Recall that this quantity is introduced in Sect. 4.3. The obtained estimates will be applied in the next sections.

Consider a nondecreasing scalar function $\mu(s)$ $(s \in [0, \eta])$, and put

$$k(z) = z + \int_0^{\eta} exp(-zs) d\mu(s) \ (z \in \mathbf{C}). \tag{6.1}$$

Obviously, $k(z)$ is the characteristic function of the scalar equation

$$\dot{y} + \int_0^\eta y(t-s)d\mu(s) = 0. \tag{6.2}$$

Lemma 4.8 *The equality*

$$\inf_{-2var(\mu)\leq\omega\leq 2var(\mu)} |k(i\omega)| = \inf_{\omega\in\mathbf{R}} |k(i\omega)|$$

is valid.

Proof Clearly, $k(0) = var(\mu)$ and

$$|k(i\omega)| \geq |\omega| - var(\mu) \geq var(\mu) \quad (|\omega| \geq 2var(\mu)).$$

This proves the lemma. □

Lemma 4.9 *Let*

$$\eta\ var(\mu) < \pi/4. \tag{6.3}$$

Then all the zeros of $k(z)$ *are in* C_- *and*

$$\inf_{\omega\in\mathbf{R}} |k(i\omega)| \geq \hat{d} > 0,$$

where

$$\hat{d} := \int_0^\eta cos(2var(\mu)\tau)d\mu(\tau). \tag{6.4}$$

Proof In this proof, for the brevity we put $v = 2var(\mu)$. Clearly,

$$|k(i\omega)|^2 = |i\omega + \int_0^\eta e^{-i\omega\tau}d\mu(\tau)|^2$$
$$= (\omega - \int_0^\eta sin\ (\tau\omega)d\mu(\tau))^2 + (\int_0^\eta cos\ (\tau\omega)d\mu(\tau))^2.$$

Hence, by the previous lemma we obtain

$$|k(i\omega)| \geq \inf_{|\omega|\leq v} |k(i\omega)| \geq \hat{d} \quad (\omega \in \mathbf{R}).$$

Furthermore, put

$$k(m, z) = m\ var(\mu)(1 + z) + (1 - m)k(z),\ \ 0 \leq m \leq 1.$$

We have

$$k(0, z) = k(z), k(1, z) = var(\mu)(1 + z) \text{ and } k(m, 0) = var(\mu).$$

Repeating the proof of the previous lemma we derive the equality

$$\inf_{\omega \in \mathbf{R}} |k(m, i\omega)| = \inf_{-v \leq \omega \leq v} |k(m, i\omega)|.$$

In addition,

$$|Re\ k(m, i\omega)| = var\ (\mu)m + (1 - m) \int_0^\eta cos(\omega\tau)d\mu.$$

Consequently,

$$|k(m, i\omega)| \geq var\ (\mu)m + (1 - m) \int_0^\eta cos(v\tau)d\mu.$$

Therefore,

$$|k(m, i\omega)| \geq var\ (\mu)m + (1 - m)\hat{d} > 0\ (\omega \in \mathbf{R}). \tag{6.5}$$

Furthermore, assume that $k(z)$ has a zero in the closed right hand plane $\overline{C}_+$. Take into account that $k(1, z) = 1 + z$ does not have zeros in $\overline{C}_+$. So $k(m_0, i\omega)$ $(\omega \in \mathbf{R})$ should have a zero for some $m_0 \in [0, 1]$, according to continuous dependence of zeros on coefficients. But due to to (6.5) this is impossible. The proof is complete. □

Remark 4.1 If

$$\mu(t) - \mu(0) > 0 \text{ for some } t < \frac{\pi}{4},$$

then

$$\int_0^\eta cos(\pi\tau)d\mu(\tau) > 0$$

and one can replace condition (6.4) by the following one:

$$\eta \; var(\mu) \le \frac{\pi}{4}.$$

Consider the scalar function

$$k_1(z) = z + \sum_{k=1}^{m} b_k e^{-h_k z} \;\; (h_k, b_k = const \ge 0).$$

The following result can be deduced from the previous lemma, but we are going to present an independent proof.

Lemma 4.10 *With the notation*

$$c = 2\sum_{k=1}^{m} b_k,$$

let

$$h_j c < \pi/2 \;\; (j = 1, \ldots, m).$$

Then all the zeros of $k_1(z)$ are in C_- and

$$\inf_{\omega \in \mathbf{R}} |k_1(i\omega)| \ge \sum_{k=1}^{m} b_k \cos\,(ch_k) > 0.$$

Proof We restrict ourselves by the case $m = 2$. In the general case the proof is similar. Put $h_1 = v$, $h_2 = h$. Introduce the function

$$f(y) = |iy + b_2 e^{-ihy} + b_1 e^{-iyv}|^2.$$

Clearly,

$$\begin{aligned} f(y) &= |iy + b_2 cos(hy) + b_1 cos(yv) - i\,(b_2 sin\,(hy) + b_1 sin\,(yv))|^2 \\ &= (b_2 cos(hy) + b_1 cos(yv))^2 + (y - b_2 sin\,(hy) - b_1 sin\,(yv))^2 \\ &= y^2 + b_2^2 + b_1^2 - 2b_2 y\, sin\,(hy) - 2b_1 y sin\,(yv) + 2b_2 b_1\, sin\,(hy)\, sin\,(yv) \\ &\quad + 2b_2 b_1 cos(yv) cos(hy). \end{aligned}$$

So

$$f(y) = y^2 + b_2^2 + b_1^2 - 2b_2 y\, sin\,(hy) - 2b_1 y sin\,(yv) + 2b_2 b_1 cos\; y(v - h). \quad (6.6)$$

But $f(0) = (b_2 + b_1)^2$ and

$$f(y) \geq (y - b_2 - b_1)^2 \geq (b_2 + b_1)^2 \quad (|y| > 2(b_2 + b_1)\ ).$$

Thus, the minimum of f is attained on $[-c, c]$ with $c = 2(b_2 + b_1)$. Then thanks to (6.6) $f(y) \geq w(y) \quad (0 \leq y \leq c)$, where

$$\begin{aligned} w(y) = y^2 + b_2^2 + b_1^2 - 2b_2 y\ sin\ (hc) \\ - 2b_1 y sin\ (vc) + 2b_2 b_1 cos\ c(h - v). \end{aligned}$$

and

$$dw(y)/dy = 2y - 2\ (b_2\ sin\ (hc) + 2b_1 sin(vc)).$$

The root of $dw(y)/dy = 0$ is $s = b_2\ sin\ (hc) + b_1 sin(vc)$. Thus

$$\begin{aligned} min_y w(y) &= s^2 + b_2^2 + b_1^2 - 2s^2 + 2b_2 b_1 cos\ c(h - v) \\ &= b_2^2 + b_1^2 - (b_2\ sin\ (hc) + b_1 sin(vc))^2 + 2b_2 b_1 cos\ c(h - v). \end{aligned}$$

Hence

$$\begin{aligned} min_y w(y) &= b_2^2 + b_1^2 - b_2^2 sin^2\ (ch) - b_1^2 sin^2\ (cv) + 2b_2 b_1 cos\ (ch) cos\ (cv) \\ &= b_2^2 cos^2\ (hc) + b_1^2 cos^2\ (vc) + 2b_2 b_1 cos(ch) cos(cv) \\ &= (b_2 cos(ch) + b_1 cos(cv))^2. \end{aligned}$$

This proves the required inequality. To prove the stability, consider the function

$$K(z, s) = z + b_1 e^{-szv} + b_2 e^{-szh} \quad (s \in [0, 1]). \tag{6.7}$$

Then all the zeros of $K(z, 0)$ are in C_- due to the just proved inequality,

$$\begin{aligned} \inf_{\omega \in \mathbf{R}} |K(i\omega, s)| &\geq b_1 cos(csv) + b_2 cos(csh) \\ &\geq b_1 cos(cv) + b_2 cos(ch) > 0 \quad (s \in [0, 1]). \end{aligned}$$

So $K(z, s)$ does not have zeros on the imaginary axis. This proves the lemma. □

4.6.2 Equations with Positive Fundamental Solutions

With constants $a \geq 0,\ b > 0$, let us consider the equation

$$\dot{u}(t) + au(t) + bu(t - h) = 0. \tag{6.8}$$

Lemma 4.11 *Let the condition*

$$hbe^{ah} < e^{-1} \tag{6.9}$$

hold. Then the Green function (the fundamental solution) $G_1(t)$ of Eq. (6.8) *is non-negative. Moreover,*

$$\int_0^\infty G_1(t)dt = \frac{1}{a+b}.$$

Proof First, we consider the Green function $G_b(t)$ of the equation

$$\dot{u} + bu(t-h) = 0 \ (b = const, \ t \geq 0). \tag{6.10}$$

Recall that G_b satisfies with the initial conditions

$$G_b(0) = 1, \ G_b(t) = 0 \ (t < 0). \tag{6.11}$$

Suppose that

$$bh < e^{-1}. \tag{6.12}$$

Since

$$\max_{\tau \geq 0} \tau e^{-\tau} = e^{-1},$$

there is a positive solution w_0 of the equation $we^{-w} = bh$. Taking $c = h^{-1}w_0$, we get a solution c of the equation $c = be^{hc}$. Put the equality $G_b(t) = e^{-ct}z(t)$ into Eq. (6.10). Then we obtain the equation

$$\dot{z} - cz + be^{hc}z(t-h) = \dot{z} + c(z(t-h) - z(t)) = 0.$$

But $z(0) = G_b(0) = 1, z(t) = 0 \ (t < 0)$. So the latter equation is equivalent to the following one:

$$z(t) = 1 + \int_0^t c[z(s) - z(s-h)]ds = 1 + c\int_0^t z(s)ds - c\int_0^{t-h} z(s)ds.$$

Consequently,

$$z(t) = 1 + c\int_{t-h}^t z(s)ds.$$

Due to the Neumann series it follows that $z(t)$ and, therefore, the Green function $G_b(t)$ of (6.10) are positive.

Furthermore, substituting the equality $u(t) = e^{-at}v(t)$ into (6.8), we get the equation

$$\dot{v}(t) + be^{ah}v(t-h) = 0.$$

According to (6.12), condition (6.9) provides the positivity of the Green function of the latter equation. Hence we get that $G_1(t) \geq 0$.

Integrating (6.8) we obtain

$$a\int_0^d G_1(t)dt+b\int_0^d G_1(t-h)dt = G_1(0)-G_1(d) = 1-G_1(d) \leq 1 \quad (d = const > 0).$$

So the integrals are uniformly bounded; thus putting $d \to \infty$, we obtain

$$a\int_0^\infty G_1(t)dt + b\int_0^\infty G_1(t-h)dt = G_1(0) = 1.$$

Take into account that

$$\int_0^\infty G_1(t-h)dt = \int_h^\infty G_1(t-h)dt = \int_0^\infty G_1(t)dt.$$

So

$$(a+b)\int_0^\infty G_1(t)dt = 1,$$

as claimed. □

Again consider Eq. (6.2), assuming that

$$e\eta \; var\;(\mu) < 1 \tag{6.13}$$

and denoting by G_+ the Green function of Eq. (6.10) with $b = var(\mu)$. Due to the previous lemma, under condition (6.13) G_+ is positive. We need the next result.

Lemma 4.12 *If condition* (6.13) *holds, then the Green function* G_μ *of Eq.* (6.2) *is nonnegative and satisfies the inequality*

$$G_\mu(t) \geq G_+(t) \geq 0 \; (t \geq 0).$$

Proof Indeed, according to the initial conditions, from (6.2) we have

$$\dot{G}_\mu(t) = 0 \ \ (0 \le t < \eta).$$

Hence, $G_\mu(t) = G_\mu(0) = 1$ $(0 \le t \le \eta)$. So for a sufficiently small $t_0 > \eta$, we can write

$$G_\mu(t) \ge 0, \ \dot{G}_\mu(t) \le 0 \ (0 \le t \le t_0).$$

Thus,

$$G_\mu(t-\eta) \ge G_\mu(t-s) \ (s \le \eta; \ 0 \le t \le t_0).$$

Hence,

$$var\ (\mu) G_\mu(t-\eta) \ge \int_0^\eta G_\mu(t-s) d\mu(s) \ (t \le t_0).$$

Taking into account (6.2), we get

$$\dot{G}_\mu(t) + var\ (\mu) G_\mu(t-\eta) = f(t)$$

with

$$f(t) = var\ (\mu) G_\mu(t-\eta) - \int_0^\eta G_\mu(t-s) d\mu(s) \ge 0 \ (0 \le t \le t_0).$$

Hence, by virtue of the Variation of Constants Formula, arrive at the relation

$$G_\mu(t) = G_+(t) + \int_0^t G_+(t-s) f(s) ds \ge G_+(t) \ (0 \le t \le t_0).$$

Extending this inequality to the whole half-line, we get the required result. □

Lemma 4.13 *Let $\mu(s)$ be a nondecreasing function satisfying the condition* (6.13). *Then the fundamental solution G_μ to Eq.* (6.2) *is nonnegative (due to the previous lemma). Moreover,*

$$\inf_{-\infty \le \omega \le \infty} \left| i\omega + \int_0^\eta exp(-i\omega s) d\mu(s) \right| = var\ (\mu)$$

and

$$\int_0^\infty G_\mu(t)dt = \frac{1}{var\ (\mu)}.$$

Proof Since $G_\mu(t) \ge 0$, from Eq. (6.2) it follows that $\dot G_\mu(t) \le 0$ for $t \ge 0$ and therefore $G_\mu(t) \le 1$. Take into account that

$$G_\mu(t) = \frac{1}{2\pi i}\int_{a-i\infty}^{a+i\infty} e^{zt}\frac{dz}{k(z)} \quad (a = const), \text{ where } k(z) = z + \int_0^\eta exp(-zs)d\mu(s)$$

and

$$\frac{1}{k(z)} = \int_0^\infty e^{-zt}G_\mu(t)dt.$$

Hence it easily follows that

$$\frac{1}{|k(i\omega)|} \le \int_0^\infty G_\mu(t)dt = \frac{1}{k(0)} \quad (\omega \in \mathbf{R}).$$

But $k(0) = var\ (\mu)$. This proves the lemma. □

Now let $\mu_0(s)$ $(s \in [0, \eta])$ be another nondecreasing scalar function with the property

$$e\eta\ var(\mu_0) \le c_0 \quad (1 \le c_0 < 2) \tag{6.14}$$

and put

$$k_0(z) = z + \int_0^\eta exp(-zs)d\mu_0(s) \ (z \in \mathbf{C}). \tag{6.15}$$

Put $\mu(s) = \mu_0(s)/c_0$. Then

$$|k_0(i\omega) - k(i\omega)| \le \int_0^\eta d(\mu_0(s) - \mu(s)) = (c_0 - 1)\ var(\mu).$$

Hence, by the previous lemma,

$$|k_0(i\omega)| \geq |k(i\omega)| - |k_0(i\omega) - k(i\omega)| \geq var(\mu) - (c_0 - 1)var(\mu)$$
$$= (2 - c_0)var(\mu) \quad (\omega \in \mathbf{R}).$$

We thus have proved the following result.

Lemma 4.14 *Let $\mu_0(s)$ $(s \in [0, \eta])$ be a nondecreasing scalar function satisfying* (6.14). *Then*

$$\inf_{\omega \in \mathbf{R}} |i\omega + \int_0^\eta exp(-i\omega s) d\mu_0(s)| \geq \frac{2 - c_0}{c_0} var(\mu_0).$$

4.6.3 Additional Lower Bounds for Quasipolynomials

Let $\mu_2(s)$ be a nondecreasing function defined on $[0, \eta]$ and

$$k_2(z) = z + ae^{-hz} + \int_0^\eta exp(-zs) d\mu_2(s) \ (z \in \mathbf{C})$$

with constants $a > 0$ and $h \geq 0$. Put

$$v_2 := 2(a + var\ (\mu_2)).$$

By Lemma 4.8,

$$\inf_{-v_2 \leq \omega \leq v_2} |k_2(i\omega)| = \inf_{\omega \in \mathbf{R}} |k_2(i\omega)|.$$

But

$$|k_2(i\omega)|^2 = (\omega - asin\ (h\omega) - \int_0^\eta sin\ (\tau\omega) d\mu_2(\tau))^2$$
$$+ (a\ cos\ (h\omega) + \int_0^\eta cos\ (\tau\omega) d\mu_2(\tau))^2.$$

Let

$$v_2 h < \pi/2 \tag{6.16}$$

and

$$d_2 := \inf_{|\omega|\le v_2} a\cos(h\omega) + \int_0^\eta \cos(\tau\omega)d\mu_2(\tau) > 0. \tag{6.17}$$

Then we obtain

$$\inf_{\omega\in\mathbf{R}} |k_2(i\omega)| \ge d_2. \tag{6.18}$$

Repeating the arguments of the proof of Lemma 4.9 we arrive at the following result.

Lemma 4.15 *Let conditions* (6.16) *and* (6.17) *hold. Then all the zeros of* $k_2(z)$ *are in* C_- *and inequality* (6.18) *is fulfilled.*

Let us point one corollary of this lemma.

Corollary 4.5 *Let the conditions* (6.16) *and*

$$a\cos(hv_2) > var(\mu_2)$$

hold. Then all the zeros of $k_2(z)$ *are in* C_- *and the inequality*

$$\inf_{\omega\in\mathbf{R}} |k_2(i\omega)| \ge a\cos(hv_2) - var(\mu_2) > 0 \tag{6.19}$$

is valid.

In particular, if

$$k_2(z) = z + a + \int_0^\eta exp(-zs)d\mu_2(s)\ (z\in\mathbf{C})$$

and

$$a > var(\mu_2), \tag{6.20}$$

then conditions (6.16) and (6.17) automatically hold, and

$$\inf_{\omega\in\mathbf{R}} |k_2(i\omega)| \ge a - var(\mu_2) > 0.$$

4.7 Autonomous Systems with One Distributed Delay

4.7.1 The General Case

In this section we illustrate our preceding results in the case of the equation

$$\dot{y}(t) + A\int_0^\eta y(t-s)d\mu = 0 \quad (t \geq 0), \tag{7.1}$$

where $A = (a_{jk})$ is a constant Hurwitzian $n \times n$-matrix and μ is a scalar nondecreasing function, again. So in the considered case we have $R_0(s) = \mu(s)A$,

$$K(z) = zI + \int_0^\eta e^{-zs} d\mu(s)A, \tag{7.2}$$

and the eigenvalues of matrix $K(z)$ with a fixed z are

$$\lambda_j(K(z)) = z + \int_0^\eta e^{-zs} d\mu(s)\lambda_j(A).$$

In addition,

$$g(B(i\omega)) = g(A\int_0^\eta e^{-i\omega s} d\mu(s)) \leq g(A)var\ (\mu) \quad (\omega \in \mathbf{R})$$

and

$$var(R_0) = \|A\|_n var\ (\mu).$$

According to Corollary 4.1 and Lemma 4.8, we have the inequality $\theta(K) \leq \theta_A$, where

$$\theta_A := \sum_{k=0}^{n-1} \frac{(g(A)var\ (\mu))^k}{\sqrt{k!}d_K^{k+1}}$$

with

$$d_K = \min_{j=1,\dots,n} \inf_{-2var(\mu)\|A\|_n \leq \omega \leq 2var(\mu)\|A\|_n} \left| \omega i + \lambda_j(A)\int_0^\eta e^{-i\omega s} d\mu(s) \right|.$$

In particular, if A is a normal matrix, then $g(A) = 0$ and $\theta_A = 1/d_K$.

Again assume that *all the characteristic values of Eq.* (7.1) *lie in* C_-.

Now Theorem 4.1 yields

Corollary 4.6 *The Cauchy operator* $\hat{G}$ *of* (7.1) *satisfies the inequality* $\|\hat{G}\|_{L^2(0,\infty)} \leq \theta_A$.

Now Corollary 4.3, formula (4.12), as well as Lemma 4.6 and Corollary 4.4 imply our next result.

Lemma 4.16 *The fundamental solution* $G(t)$ *of Eq.* (7.1) *is subject to the inequalities*

$$\|G\|_{L^2(0,\infty)} \leq (\int_0^\infty N_2^2(G(t))dt)^{1/2} \leq W(A,\mu), \tag{7.3}$$

where

$$W(A,\mu) := \sqrt{\frac{n}{\pi}(\theta_A^2 \ \|A\|_n var(\mu) + 2\theta_A)}.$$

Moreover,

$$\|\dot{G}\|_{L^2(0,\infty)} \leq \|A\|_n var(\mu) \ W(A,\mu), \tag{7.4}$$

$$\|G\|_{C(0,\infty)} \leq W(A,\mu)\sqrt{2\|A\|_n \ var(\mu)} \tag{7.5}$$

and

$$\|G\|_{L^1(0,\infty)} \leq \theta_A^{3/2}\sqrt{3n\|A\|_n var(\mu)}\,(1 + \|A\|_n \ vd \ (\mu)), \tag{7.6}$$

where $vd\ (\mu) = \int_0^\eta s d\mu(s)$.

Since $\|\hat{G}\|_{L^p(0,\infty)} \leq \|G\|_{L^1(0,\infty)}$, we arrive at the main result of the present section.

Theorem 4.3 *The Cauchy operator* $\hat{G}$ *of* (7.1) *satisfies the inequality*

$$\|\hat{G}\|_{L^p(0,\infty)} \leq \theta_A^{3/2}\sqrt{3n\|A\|_n var(\mu)}\,(1 + \|A\|_n \ vd \ (\mu)).$$

Besides, $\|\hat{G}\|_{C(0,\infty)} = \|\hat{G}\|_{L^\infty(0,\infty)}$.

4.7.2 Application of Lemma 4.9

Clearly, d_K can be directly calculated. In addition, by Lemma 4.9 we get the following result.

Lemma 4.17 *Let all the eigenvalues of A be real and positive:*

$$0 < \lambda_1(A) \le \cdots \le \lambda_n(A) \tag{7.7}$$

and let

$$\eta \; var(\mu)\lambda_n(A) < \frac{\pi}{4}. \tag{7.8}$$

Then all the characteristic values of Eq. (7.1) *are in* C_- *and*

$$d_K \ge d(A, \mu), \tag{7.9}$$

where

$$d(A, \mu) := \min_j \lambda_j(A) \int_0^\eta cos\left(2\tau\lambda_j(A) var\ (\mu)\right) d\mu(\tau) > 0.$$

Obviously,

$$d(A, \mu) > \lambda_1(A) \int_0^\eta cos\,(2\tau\lambda_n(A) var\ (\mu))\, d\mu(\tau) > 0.$$

From the latter lemma, we obtain

$$\theta_A \le \theta(A, \mu), \text{ where } \theta(A, \mu) := \sum_{k=0}^{n-1} \frac{(g(A) var\ (\mu))^k}{\sqrt{k!} d^{k+1}(A, \mu)}.$$

So due to Corollary 4.6, we arrive at the following result.

Corollary 4.7 *Let conditions* (7.7) *and* (7.8) *hold. Then all the characteristic values of Eq.* (7.1) *are in* C_- *and* $\|\hat{G}\|_{L^2(0,\infty)} \le \theta(A, \mu)$. *Moreover, Lemma 4.16 and Theorem 4.3 are valid with* $\theta(A, \mu)$ *instead of* θ_A.

In particular, if (7.1) takes the form

$$\dot{y}(t) + Ay(t - h) = 0 \quad (h = const > 0;\ t \ge 0), \tag{7.10}$$

then the eigenvalues of $K(z)$ are $\lambda_j(K(z)) = z + e^{-zh}\lambda_j(A)$. In addition, $var(\mu) = 1$, $vd(\mu) = h$. According to (7.7), condition (7.8) takes the form

$$h\lambda_n(A) < \pi/4. \tag{7.11}$$

Besides, in this case we obtain the inequalities

$$d(A, \mu) \geq \min_j \lambda_j(A)cos(2\lambda_j(A)h) \geq \lambda_1(A)cos(2\lambda_n(A)h). \tag{7.12}$$

4.7.3 Application of Lemma 4.13

Furthermore, under condition (7.7), instead of condition (7.8) assume that

$$e\eta\ var(\mu)\ \lambda_n(A) < 1. \tag{7.13}$$

Then according to Lemma 4.13 we can write

$$\inf_{\omega \in \mathbf{R}} |\lambda_j(K(i\omega))| = \inf_{\omega \in \mathbf{R}} \left| i\omega + \lambda_j(A) \int_0^\eta exp(-i\omega s)d\mu(s) \right| = \lambda_j(A)var(\mu).$$

Hence,

$$d_K = var(\mu)\lambda_1(A), \tag{7.14}$$

and therefore, $\theta_A \leq \hat{\theta}_A$, where

$$\hat{\theta}_A := \frac{1}{var\ (\mu)} \sum_{k=0}^{n-1} \frac{g^k(A)}{\sqrt{k!}\lambda_1^{k+1}(A)}.$$

So due to Corollary 4.6, we arrive at the following result.

Corollary 4.8 *Let conditions* (7.7) *and* (7.13) *hold. Then all the characteristic values of Eq.* (7.1) *are in* C_- *and* $\|\hat{G}\|_{L^2(0,\infty)} \leq \hat{\theta}_A$. *Moreover, Lemma 4.16 and Theorem 4.3 are valid with* $\theta_A = \hat{\theta}_A$.

If A is a positive definite Hermitian $n \times n$-matrix, then the previous corollary can be improved. Indeed, rewrite (7.1) as the system in the basis of the eigenvectors of A:

$$\dot{y}_j(t) + \lambda_j(A) \int_0^\eta y_j(t-s)d\mu(s) = 0 \ \ (t > 0; j = 1, \ldots, n). \tag{7.15}$$

Again assume that condition (7.13) holds. Then due to Lemma 4.13, we can assert that the Green function (the fundamental solution) $G_j(t)$ to Eq. (7.15) for each j is nonnegative and

$$\int_0^\infty G_j(t)dt = \frac{1}{var(\mu)\lambda_j(A)}.$$

Besides, the fundamental solution $G(t)$ to the vector Eq. (7.1) is the diagonal matrix valued function with the diagonal entries $G_j(t)$. Put

$$x_j(t) = \int_0^t G_j(t-s)f_j(s)ds \tag{7.16}$$

for a scalar continuous function f_j. Then

$$\sup_t \left(\sum_{k=1}^n |x_k(t)|^2\right)^{1/2} = \sup_t \left(\sum_{j=1}^n |\int_0^t G_j(t-s)f_j(s)ds|^2\right)^{1/2}$$
$$\le \sqrt{n} \sup_{t,k} |f_k(t)| \int_0^\infty G_k(t)dt \le \sqrt{n}\|f\|_{C(0,\infty)} \sup_k \frac{1}{var(\mu)\lambda_k(A)}.$$

We thus have proved the following result

Lemma 4.18 *Let A be a positive definite Hermitian $n \times n$-matrix and conditions* (7.7) *and* (7.13) *hold. Then the Cauchy operator to* (7.1) *satisfies the inequality*

$$\|\hat{G}\|_{C(0,\infty)} \le \frac{\sqrt{n}}{var(\mu)\lambda_1(A)}.$$

4.7.4 Equations with Diagonalizable Matrices

Let A be a diagonalizable matrix. Then there is an invertible matrix T and a normal matrix S, such that $T^{-1}AT = S$. Besides, for the characteristic matrix function of (7.1) we can write

$$T^{-1}K(z)T = K_S(z), \text{ where } K_S(z) = zI + \int_0^\eta e^{-zs}d\mu(s)S$$

and therefore,

$$\|K^{-1}(z)\|_n \le \kappa(A,T)\|K_S^{-1}(z)\|_n$$

where $\kappa(A, T) = \|T^{-1}\|_n \|T\|_n$. Since $g(S) = 0$, and the eigenvalues of K and K_S coincide, we have $d_K = d_S$, and

$$\theta_S = \sup_{-2var(\mu)\|S\|_n \le \omega \le 2var(\mu)\|S\|_n} \|K_S^{-1}(z)\|_n \le \frac{1}{d_K}.$$

Now Corollary 4.6 implies the following result.

Corollary 4.9 *Assume that all the characteristic values of Eq.* (7.1) *lie in* C_- *and* A *is a diagonalizable matrix. Then the Cauchy operator* $\hat{G}$ *of* (7.1) *satisfies the inequality*

$$\|\hat{G}\|_{L^2(0,\infty)} \le \frac{\kappa(A, T)}{d_K}.$$

Moreover, Lemma 4.16 and Theorem 4.3 are valid with $\frac{\kappa(A,T)}{d_K}$ *instead of* θ_A.

If, in addition, the eigenvalues of a diagonalizable matrix A are real, and conditions (7.7) and (7.8) hold, then according to (7.9), Eq. (7.1) is stable and one can replace d_K by $d(A, \mu)$.

Similarly, according to Lemma 4.18, we obtain the following result

Lemma 4.19 *Let be* A *a diagonalizable* $n \times n$*-matrix with positive eigenvalues and condition* (7.13) *hold. Then Eq.* (7.1) *is exponentially stable. In addition, the Cauchy operator to* (7.1) *satisfies the inequality*

$$\|\hat{G}\|_{C(0,\infty)} \le \frac{\kappa(A, T)\sqrt{n}}{var(\mu) \min_k \lambda_k(A)}.$$

4.8 Solution Estimates Via Determinants

Again consider Eq. (1.1). As it was shown in Sect. 2.3,

$$\|A^{-1}\|_n \le \frac{N_2^{n-1}(A)}{(n-1)^{(n-1)/2}|det\ (A)|} \quad (n \ge 2),$$

for an invertible $n \times n$-matrix A. Hence, for any regular point z of the characteristic function of Eq. (1.1), one has

$$\|K^{-1}(z)\|_n \le \frac{N_2^{n-1}(K(z))}{(n-1)^{(n-1)/2}|det\ (K(z))|} \quad (n \ge 2). \tag{8.1}$$

Recall that it is assumed that all the characteristic values of (1.1) are in the open left half-plane C_-. Our next result directly follows from inequality (8.1).

Corollary 4.10 *One has $\theta(K) \le \theta_{det}(K)$, where*

$$\theta_{det}(K) := \sup_{-2\,var(R_0)\le\omega\le 2\,var(R_0)} \frac{N_2^{n-1}(K(i\omega))}{(n-1)^{(n-1)/2}|det\ (K(i\omega))|},$$

and therefore (due to Theorem 4.1), the Cauchy operator to Eq. (1.1) *satisfies the inequality*

$$\|\hat{G}\|_{L^2(0,\infty)} \le \theta_{det}(K).$$

Note that $\theta_{det}(K)$ is more convenient for the calculations than $\Gamma_0(K)$, if n is small enough.

We have $W(K) \le W_{det}(K)$, where

$$W_{det}(K) := \sqrt{\frac{n}{\pi}(\theta^2_{det}(K)var(R_0) + 2\theta_{det}(K))}.$$

According to (4.8)

$$W^2_{det}(K) \le \theta^2_{det}(K)\frac{3n\ var(R_0)}{\pi} \le \theta^2_{det}(K)n\ var(R_0).$$

By Corollary 4.3 we arrive at the following result.

Lemma 4.20 *The inequality*

$$\int_0^\infty N_2^2(G(t))dt \le W^2_{det}(K)$$

is valid.

In addition, inequality (4.11) and Lemma 4.6 imply our next result.

Lemma 4.21 *The inequalities*

$$\|\dot{G}\|_{L^2(0,\infty)} \le W_{det}(K)var(R_0)$$

and

$$\|G\|_{C(0,\infty)} \le a_{det}(K), \tag{8.2}$$

hold, where

$$a_{det}(K) := \sqrt{2var(R_0)}W_{det}(K).$$

Moreover, we have $\zeta(K) \le \zeta_{det}(K)$, where

$$\zeta_{det}(K) := \theta_{det}^{3/2}(K)\sqrt{3n\, var(R_0)\,(1 + vd(R_0))}.$$

Now Theorem 4.2 implies

Corollary 4.11 *The Cauchy operator to Eq.* (1.1) *satisfies the inequality*

$$\|\hat{G}\|_{L^p(0,\infty)} \le \zeta_{det}(K) \quad (p \ge 1).$$

Besides, $\|\hat{G}\|_{C(0,\infty)} = \|\hat{G}\|_{L^\infty(0,\infty)}$.

Furthermore, if the entries $r_{jk}(\tau)$ of $R_0(\tau)$ have the following properties: r_{jj} are non-increasing and $r_{jk}(\tau) \equiv 0$ $(j > k)$, that is, R_0 is upper triangular; then, clearly,

$$det\ K(z) = \prod_{k=1}^{n} \left(z - \int_0^\eta e^{-zs} dr_{kk}(s) \right). \tag{8.3}$$

If, in addition

$$\eta\ var\ r_{jj} \le \frac{\pi}{4}, j = 1, \ldots, n, \tag{8.4}$$

then by Lemma 4.9 all the zeros of $det\ K(z)$ are in C_- and

$$|det\ K(i\omega)| \ge \prod_{k=1}^{n} \hat{d}_{kk} > 0, \tag{8.5}$$

where

$$\hat{d}_{jj} = \int_0^\eta cos(2var(r_{jj})\tau)\, dr_{jj}(\tau).$$

About the determinants of perturbations of triangular matrices see Sect. 2.10.

4.9 Diagonally Dominant Differential Delay Systems

Again $r_{jk}(s)$ $(j, k = 1, \ldots, n)$ are the entries of $R_0(s)$. In addition $r_{jj}(s)$ are non-increasing. Put

$$\xi_j = \sum_{k=1, k \ne j}^{n} var(r_{jk}).$$

Lemma 4.22 *Assume that all the zeros of the functions*

$$k_j(z) := z - \int_0^\eta e^{-zs} dr_{jj}(s),\ j = 1, \ldots, n,$$

are in C_- and, in addition,

$$|k_j(i\omega)| > \xi_j \ \ (\omega \in \mathbf{R};\ |\omega| \le 2var(r_{jj});\ j = 1, \ldots, n), \tag{9.1}$$

then Eq. (1.1) *is exponentially stable. Moreover,*

$$|det\ (K(i\omega))| \ge \prod_{j=1}^{n} (|k_j(i\omega)| - \xi_j) \ \ (\omega \in \mathbf{R}). \tag{9.2}$$

Proof As it was shown in Sect. 2.10, for any $n \times n$-matrix

$$M = (m_{jk})_{j,k=1}^{n},$$

the inequality

$$|det\ M| \ge \prod_{j=1}^{n} (|m_{jj}| - \sum_{i=1, i \ne j}^{n} |m_{ji}|)$$

is valid, provided

$$|m_{jj}| > \sum_{i=1, i \ne j}^{n} |m_{ji}| \ \ (j = 1, \ldots, n).$$

Hence, with the notation

$$k_{jm}(z) = \int_0^\eta e^{-zs} dr_{jm}(s) (m \ne j)$$

for a $z \in \mathbf{C}$ we have the inequality

$$|det\ K(z)| \ge \prod_{j=1}^{n} (|k_j(z)| - \sum_{m=1, m \ne j}^{n} |k_{jm}(z)|),$$

if the condition

$$|k_j(z)| > \sum_{m=1, m \ne j}^{n} |k_{jm}(z)|.$$

is fulfilled. It can be directly checked that

$$|k_{jk}(z)| \le \sup_{\omega\in\mathbf{R}} |k_{jk}(i\omega)| \le var(\mu_{jk}) \ \ (j \ne k;\ \ Re\, z \ge 0).$$

Due to the minimum moduli principle for analytic functions we have

$$k_j(z) \ge \inf_{\omega\in\mathbf{R}} k_j(i\omega) \text{ for } Re\, z \ge 0.$$

Assume that (9.1) holds, then

$$|k_j(z)| > \inf_{\omega\in\mathbf{R}} |k_j(i\omega)| > \sum_{m=1,m\ne j}^{n} var(\mu_{jk}) > \sum_{m=1,m\ne j}^{n} |k_{jm}(z)| \ \ (Re\, z \ge 0).$$

So (9.2) is valid, as claimed. □

If condition (8.4) holds, then by Lemma 4.9 we obtain

$$|k_j(i\omega)| \ge \hat{d}_{jj} > 0,$$

where $\hat{d}_{jj}$ are defined in the previous section. So due to the previous lemma we arrive at the following result.

Corollary 4.12 *If the conditions* (8.4) *and*

$$\hat{d}_{jj} > \xi_j \ \ (j = 1, \dots, n)$$

hold, then Eq. (1.1) *is exponentially stable. Moreover,*

$$|det\ (K(i\omega))| \ge \prod_{j=1}^{n} (\hat{d}_{jj} - \xi_j) \ \ (\omega \in \mathbf{R}).$$

Now one can apply the results of the previous section.

4.10 Time-Variant Equations Close to Ordinary Differential Ones

Let $R(t,\tau) = (r_{jk}(t,\tau))_{j,k=1}^{n}$ be an $n \times n$-matrix-valued function defined on $[0,\infty) \times [0,\eta]$ $(0 < \eta < \infty)$, which is piece-wise continuous in t for each τ, whose entries have bounded variations and finite numbers of jumps in τ.

Consider the equation

$$\dot{x}(t) = A(t)x(t) + \int_0^\eta d_\tau R(t,\tau)x(t-\tau) + f(t) \tag{10.1}$$

where $A(t)$ is a piece-wise continuous matrix valued function, $f \in C(0,\infty)$ or $f \in L^1(0,\infty)$. In this section we establish norm estimates for the Cauchy operator $\hat{G}$ to Eq. (10.1), which *do not depend on delay*.

Introduce on $L^1(-\eta,\infty)$ the operator

$$Ew(t) = \int_0^\eta d_\tau R(t,\tau)w(t-\tau) \quad (t \geq 0).$$

It is assumed that

$$v_{jk} = \sup_{t\geq 0} var(r_{jk}(t,.)) < \infty. \tag{10.2}$$

By Lemma 1.10, there is a constant

$$V_{L^1}(R) \leq \sum_{j=1}^n \sqrt{\sum_{k=1}^n v_{jk}^2},$$

such that

$$\|Ew\|_{L^1(0,\infty)} \leq V_{L^1}(R)\|w\|_{L^1(-\eta,\infty)} \quad (w \in L^1(-\eta,\infty)). \tag{10.3}$$

For instance, if (10.1) takes the form

$$\dot{y}(t) = A(t)y + \int_0^\eta B(t,s)y(t-s)ds + \sum_{k=1}^m B_k(t)y(t-\tau_k) \quad (t \geq 0;\ m < \infty), \tag{10.4}$$

where $0 \leq \tau_0 < \tau_1 < \cdots < \tau_m \leq \eta$ are constants, $B_k(t)$ are piece-wise continuous matrices and $B(t,s)$ is a matrix function integrable in s on $[0,\eta]$, then (10.2) holds, provided

$$\hat{V}_{L^1}(R) := \left(\int_0^\eta \sup_{t\geq 0}\|B(t,s)\|_n ds + \sum_{k=1}^m \sup_{t\geq 0}\|B_k(t)\|_n\right) < \infty.$$

Indeed,

$$\int_0^\infty \|Ef(t)\|_n dt \le \int_0^\infty \left(\int_0^\eta \|B(t,s)f(t-s)\|_n ds + \sum_{k=1}^m \|B_k(t)f(t-\tau_k)\|_n \right) dt$$

$$\le \int_0^\eta \sup_\tau \|B(\tau,s)\|_n \int_0^\infty \|f(t-s)\|_n ds\, dt$$

$$+ \sum_{k=1}^m \sup_\tau \|B_k(\tau)\|_n \int_0^\infty \|f(t-\tau_k)\|_n dt.$$

Consequently,

$$\int_0^\infty \|Ef(t)\|_n dt \le \hat{V}_{L^1}(R) \left(\max_{0\le s\le\eta} \int_0^\infty \|f(t-s)\|_n dt + \int_0^\infty \sum_{k=1}^m \|f(t-\tau_k)\|_n dt \right).$$

But

$$\max_{0\le s\le\eta} \int_0^\infty \|f(t-s)\|_n dt = \int_{-\eta}^\infty \|f(t)\|_n dt.$$

Thus, in the case of Eq. (10.4), condition (10.3) holds with $V_{L^1}(R) = \hat{V}_{L^1}(R)$.

Recall that the evolution operator $U(t,s)$ $(t \ge s \ge 0)$ of the equation

$$\dot{y} = A(t)y \ (t > 0) \tag{10.5}$$

is defined by $U(t,s)y(s) = y(t)$ for a solution $y(t)$ of (10.5).

Lemma 4.23 *Let condition* (10.3) *hold and let the evolution operator* $U(t,s)$ $(t \ge s \ge 0)$ *of* (10.5) *satisfy the inequality*

$$\nu_1 := \sup_{s\ge 0} \int_s^\infty \|U(t,s)\|_n dt < \frac{1}{V_{L^1}(R)}. \tag{10.6}$$

Then the Cauchy operator to Eq. (10.1) *satisfies the inequality*

$$\|\hat{G}\|_{L^1(0,\infty)} \le \frac{\nu_1}{1-\nu_1 q_1}. \tag{10.7}$$

Proof Equation (10.1) with the zero initial condition is equivalent to the following one:

$$x(t) = \int_0^t U(t, s)(Ex(s) + f(s))ds. \tag{10.8}$$

So

$$\|x(t)\|_n \le \int_0^t \|U(t, s)\|_n (\|Ex(s)\|_n + \|f(s)\|_n)ds.$$

Integrating this inequality, we obtain

$$\int_0^{t_0} \|x(t)\|_n dt \le \int_0^{t_0}\int_0^t \|U(t, s)\|_n (\|Ex(s)\|_n + \|f(s)\|_n)ds\, dt \quad (0 < t_0 < \infty).$$

Take into account that

$$\int_0^{t_0}\int_0^t \|U(t, s)\|_n \|f(s)\|_n ds\, dt = \int_0^{t_0} \|f(s)\|_n \int_s^{t_0} \|U(t, s)\|_n dt\, ds \le \nu_1 \|f\|_{L^1(0,\infty)}.$$

In addition,

$$\int_0^{t_0}\int_0^t \|U(t, s)\|_n \|Ex(s)\|_n ds\, dt$$
$$= \int_0^{t_0} \|Ex(s)\|_n \int_s^{t_0} \|U(t, s)\|_n dt\, ds \le V_{L^1}(R)\nu_1 \int_0^{t_0} \|x(s)\|_n ds.$$

Thus,

$$\int_0^{t_0} \|x(s)\|_n ds \le \nu_1 V_{L^1}(R) \int_0^{t_0} \|x(s)\|_n ds + \nu_1 \|f\|_{L^1(0,\infty)}.$$

Hence,

$$\int_0^{t_0} \|x(s)\|_n ds \le \frac{\nu_1 \|f\|_{L^1(0,t_0)}}{1 - \nu_1 V_{L^1}(R)}.$$

Now letting $t_0 \to \infty$, we arrive at the required result. □

Now consider the Eq. (10.1) in space $C(0, \infty)$. By Lemma 1.10, under condition (10.2) there is a constant

$$V_C(R) \leq \sqrt{n}\| \left(v_{jk}\right)_{j,k=1}^n \|_n,$$

such that

$$\|Ew\|_{C(0,\infty)} \leq V_C(R)\|w\|_{C(-\eta,\infty)} \quad (w \in C(-\eta, \infty)). \tag{10.9}$$

For instance, if (10.1) takes the form

$$\dot{y}(t) = A(t)y(t) + \int_0^\eta B(t, s)y(t-s)ds + \sum_{k=1}^m B_k(t)y(t - h_k(t)) \quad (t \geq 0;\ m < \infty), \tag{10.10}$$

where $0 \leq h_1(t), h_2(t), \ldots, h_m(t) \leq \eta$ are continuous functions, $B(t, s)$, $B_k(t)$ and $\hat{V}_{L^1}(R) < \infty$ are the same as above in this section, then

$$\sup_{t\geq 0} \|Ew(t)\|_n \leq \sup_{t\geq 0} \left(\int_0^\eta \|B(t, s)w(t-s)\|_n ds + \sum_{k=1}^m \|B_k(t)w(t - h_k(t))\|_n \right).$$

Hence, under the condition $\hat{V}_{L^1}(R) < \infty$, we easily obtain inequality (10.9) with $V_C(R) = \hat{V}_{L^1}(R)$.

Lemma 4.24 *Let condition* (10.9) *hold and the evolution operator* $U(t, s)$ $(t \geq s \geq 0)$ *of Eq.* (10.5) *satisfy the inequality*

$$\nu_\infty := \sup_{t\geq 0} \int_0^t \|U(t, s)\|_n ds < \frac{1}{V_C(R)}. \tag{10.11}$$

Then the Cauchy operator to Eq. (10.1) *satisfies the inequality*

$$\|\hat{G}\|_{C(0,\infty)} \leq \frac{\nu_\infty}{1 - \nu_\infty V_C(R)}.$$

The proof of this lemma is similar to the proof of Lemma 4.23.

Let $(.,.)_{C^n}$ be the scalar product in $\mathbf{C}^n$ and

$$((A(t) + A^*(t))h, h)_{C^n} \leq -2c(t)(h, h)_{C^n} \quad (h \in \mathbf{C}^n, t \geq 0)$$

with a positive piece-wise continuous function $c(t)$ having the property

$$\hat{\nu}_1 := \sup_{s\geq 0} \int_s^\infty e^{-\int_s^t c(t_1)dt_1} dt < \infty.$$

Then we easily have the inequalities

$$\|U(t,s)\|_n \leq e^{-\int_s^t c(t_1)dt_1}$$

and $\nu_1 \leq \hat{\nu}_1$. For instance, if $c_0 := \inf_t c(t) > 0$, then

$$\hat{\nu}_1 \leq \sup_{s\geq 0} \int_s^\infty e^{-c_0(t-s)} dt = \frac{1}{c_0}.$$

Let $A(t) \equiv A$ be a constant Hurwitzian matrix. Then

$$U(t,s) = e^{A(t-s)} \quad (t \geq s \geq 0).$$

In this case

$$\nu_1 = \nu_\infty = \|e^{At}\|_{L^1(0,\infty)}.$$

Applying Corollary 2.5, we have

$$\|e^{At}\|_n \leq e^{\alpha(A)t} \sum_{k=0}^{n-1} \frac{g^k(A)t^k}{(k!)^{3/2}} \quad (t \geq 0).$$

Recall that

$$g(A) = (N_2^2(A) - \sum_{k=1}^n |\lambda_k(A)|^2)^{1/2} \leq \sqrt{2}N_2(A_I),$$

and $\alpha(A) = \max_k Re\, \lambda_k(A) < 0$; $\lambda_k(A)$ $(k = 1, \ldots, n)$ are the eigenvalues of A, $N_2(A)$ is the Hilbert-Schmidt norm of A and $A_I = (A - A^*)/2i$. Thus

$$\|e^{At}\|_{L^1(0,\infty)} \leq \nu_A,$$

where

$$\nu_A := \sum_{k=0}^{n-1} \frac{g^k(A)}{\sqrt{k!}|\alpha(A)|^{k+1}}.$$

4.11 Equations with Slowly Varying Coefficients

In this subsection we investigate Eq. (10.1) under the condition

$$\|A(t) - A(s)\|_n \le q_0|t-s| \ (t, s \ge 0), \tag{11.1}$$

applying the freezing method.

Theorem 4.4 *Let the conditions* (11.1),

$$\nu_A := \sup_{s\ge 0} \int_0^\infty \|e^{A(s)t}\|_n dt < \frac{1}{V_C(R)} \tag{11.2}$$

and

$$\hat{\chi}_0 := \sup_{s\ge 0} \int_0^\infty t\|e^{A(s)t}\|_n dt < \frac{1-\nu_A V_C(R)}{q_0} \tag{11.3}$$

hold. Then the Cauchy operator $\hat{G}_1$ *of Eq.* (11.1) *satisfies the inequality*

$$\|\hat{G}_1\|_{C(0,\infty)} \le \tau(A,E) := \frac{\nu_A}{1 - q_0\hat{\chi}_0 - V_C(R)\nu_A}.$$

To prove this theorem put

$$\xi_A := \sup_{f\in C(0,\infty)} \frac{1}{\|f\|_{C(0,\infty)}} \sup_{t\ge 0} \left\| \int_0^t U(t,s)f(s)ds \right\|_n .$$

We need the following technical lemma.

Lemma 4.25 *Let the condition*

$$\xi_A V_C(R) < 1 \tag{11.4}$$

hold. Then the Cauchy operator $\hat{G}_1$ *of Eq.* (10.1) *satisfies the inequality*

$$\|\hat{G}_1\|_{C(0,\infty)} \le \frac{\xi_A}{1 - V_C(R)\xi_A}.$$

Proof Equation (10.1) is equivalent to the following one:

$$x(t) = \int_0^t U(t,s)(Ex(s) + f(s))ds.$$

Hence,

$$\|x\|_{C(0,\infty)} \leq \xi_A(\|Ex\|_{C(0,\infty)} + \|f\|_{C(0,\infty)}).$$

We thus we arrive at the inequality

$$\|x\|_{C(0,\infty)} \leq \xi_A(V_C(R)\|x\|_{C(0,\infty)} + \|f\|_{C(0,\infty)}).$$

Now condition (11.4) ensures the required result. □

Lemma 4.26 *Let conditions* (11.1) *and* (11.3) *hold. Then*

$$\xi_A < \frac{\nu_A}{1 - q_0\hat{\chi}_0}.$$

Proof Consider the equation

$$\dot{x}(t) = A(t)x(t) + f(t) \tag{11.5}$$

with the zero initial condition $x(0) = 0$. Rewrite it as

$$\dot{x}(t) = A(s)x(t) + (A(t) - A(s))x(t) + f(t).$$

Hence

$$x(t) = \int_0^t e^{A(s)(t-t_1)}[(A(t_1) - A(s))x(t_1) + f(t_1)]dt_1.$$

Take $s = t$. Then

$$\|x(t)\|_n \leq \int_0^t \|e^{A(t)(t-t_1)}\| \|(A(t_1) - A(t))x(t_1)\|dt_1 + c_0,$$

where

$$c_0 := \sup_{s,t} \int_0^t \|e^{A(s)(t-t_1)}\|_n \|f(t_1)\|_n dt_1$$

$$\leq \|f\|_{C(0,\infty)} \sup_s \int_0^\infty \|e^{A(s)t_1}\|_n dt_1 \leq \nu_A \|f\|_{C(0,\infty)}.$$

Thus, for any $T < \infty$, we get

$$\sup_{t \le T} \|x(t)\|_n \le c_0 + q_0 \sup_{t \le T} \|x(t)\|_n \int_0^T \|e^{A(t)(T-t_1)}\| |t_1 - T| dt_1$$

$$\le c_0 + q_0 \sup_{t \le T} \|x(t)\|_n \int_0^T \|e^{A(t)u}\| u du \le c_0 + q_0 \hat{\chi}_0 \sup_{t \le T} \|x(t)\|_n.$$

By (11.3), we have $q_0 \hat{\chi}_0 < 1$. So

$$\|x\|_{C(0,T)} \le \frac{c_0}{1 - q_0 \hat{\chi}_0} = \frac{\nu_A \|f\|_{C(0,\infty)}}{1 - q_0 \hat{\chi}_0}.$$

Hence, letting $T \to \infty$, we get

$$\|x\|_{C(0,\infty)} \le \frac{c_0}{1 - q_0 \hat{\chi}_0} = \frac{\nu_A \|f\|_{C(0,\infty)}}{1 - q_0 \hat{\chi}_0}.$$

This proves the lemma. □

Proof of Theorem 4.4 The required result at once follows from Lemma 4.25 and the previous lemma. □

4.12 Time-Variant Equations with Small Delays

Again consider the equation

$$\dot{y}(t) = \int_0^\eta d_\tau R(t, \tau) y(t - \tau) \quad (t \ge 0), \tag{12.1}$$

where $R(t, \tau) = (r_{jk}(t, \tau))$ is the same as in Sect. 4.10. In particular, condition (10.2) holds.

For instance, (12.1) can take the form

$$\dot{y}(t) = \int_0^\eta B(t, s) y(t - s) ds + \sum_{k=1}^m B_k(t) y(t - \tau_k) \quad (t \ge 0;\ m < \infty), \tag{12.2}$$

where $B(t, s)$, τ_k and $B_k(t)$ are the same as in Sect. 4.10. In $C(-\eta, \infty)$ introduce the operator

$$\hat{E}_d f(t) := \int_0^\eta d_\tau R(t,\tau)[f(t-\tau) - f(t)] \quad (t \ge 0).$$

Assume that there is a constant $\tilde{V}_d(R)$, such that

$$\|\hat{E}_d f\|_{C(0,T)} \le \tilde{V}_d(R) \|\dot{f}\|_{C(-\eta,T)} \quad (T > 0) \tag{12.3}$$

for any $f \in C(-\eta, T)$ with $\dot{f} \in C(-\eta, T)$.

In the case of Eq. (12.2) we have

$$\hat{E}_d f(t) = \int_0^\eta B(t,\tau)[f(t-\tau) - f(t)]d\tau + \sum_{k=1}^m B_k(t)[f(t-\tau_k) - f(t)].$$

Since

$$\|f(t-\tau) - f(t)\|_{C(0,T)} \le \tau \|\dot{f}\|_{C(-\eta,T)},$$

we easily obtain

$$\|\hat{E}_d f\|_{C(0,T)} \le \sup_t \left(\int_0^\eta \tau \|B(t,\tau)\|_n d\tau + \sum_{k=1}^m \|B_k(t)\|_n \tau_k \right) \|\dot{f}\|_{C(-\eta,T)}.$$

That is, for Eq. (12.2), condition (12.3) holds with

$$\tilde{V}_d(R) \le \sup_t \left(\int_0^\eta \tau \|B(t,\tau)\|_n d\tau + \sum_{k=1}^m \tau_k \|B_k(t)\|_n \right). \tag{12.4}$$

Put

$$A(t) = R(t,\eta) - R(t,0) \tag{12.5}$$

and assume that the ordinary differential Eq. (10.5) is asymptotically stable. Recall that $U(t,s)$ is the evolution operator of Eq. (10.5) and again put

$$\nu_\infty = \sup_{t \ge 0} \int_0^t \|U(t,s)\|_n ds.$$

Theorem 4.5 *Under conditions* (10.2) *and* (12.3), *let* $A(t)$ *be defined by* (12.5) *and*

$$\psi_R := \tilde{V}_d(R) \left(\nu_\infty \|A(.)\|_{C(0,\infty)} + 1 \right) < 1. \tag{12.6}$$

Then the Cauchy operator to Eq. (12.1) is subject to the inequality

$$\|\hat{G}\|_{C(0,\infty)} \leq \frac{\nu_\infty}{1 - \psi_R}.$$

Proof We need the non-homogeneous problem

$$\dot{x}(t) = Ex(t) + f \quad (t \geq 0;\ f \in C(0, \infty)), \tag{12.7}$$

with the zero initial condition

$$x(t) = 0, t \leq 0.$$

Observe that

$$\int_0^\eta d_\tau R(t,\tau)x(t-\tau) = \int_0^\eta d_\tau R(t,\tau)x(t) + \int_0^\eta d_\tau R(t,\tau)(x(t-\tau) - x(t))$$

$$= (R(t,\eta) - R(t,0))x(t) + \int_0^\eta d_\tau R(t,\tau)(x(t-\tau) - x(t))$$

$$= A(t)x(t) + \hat{E}_d x(t).$$

So we can rewrite Eq. (12.7) as

$$\dot{x}(t) = A(t)x(t) + \hat{E}_d x(t) + f(t) \quad (t \geq 0). \tag{12.8}$$

Consequently,

$$x(t) = \int_0^t U(t,s)\hat{E}_d x(s)ds + f_1(t), \tag{12.9}$$

where

$$f_1(t) = \int_0^t U(t,s)f(s)ds.$$

Differentiating (12.9), we get

$$\dot{x}(t) = \int_0^t A(t)U(t,s)\hat{E}_d x(s)ds + \hat{E}_d x(t) + A(t)f_1(t) + f(t).$$

For the brevity put $|x|_T = \|x\|_{C(0,T)}$ for a finite T. By condition (12.3) we have

$$|\dot{x}|_T \le c_0 + \tilde{V}_d(R)|\dot{x}|_T \left(\sup_{t\ge 0} \int_0^t \|A(t)U(t,s)\|_n ds + 1 \right),$$

where

$$c_0 := |A(.)f_1|_T + |f|_T.$$

So

$$|\dot{x}|_T \le c_0 + \psi_R |\dot{x}|_T.$$

Now condition (12.6) implies

$$|\dot{x}|_T \le \frac{c_0}{1-\psi_R}.$$

But due to (12.9) and (12.3) the following inequalities are true:

$$|x|_T \le \nu_\infty \tilde{V}_d(R)|\dot{x}|_T + |f_1|_T \le \nu_\infty \tilde{V}_d(R)\frac{c_0}{1-\psi_R} + |f_1|_T.$$

Take into account that $|f_1|_T \le \nu_\infty |f|_T$ and

$$c_0 \le |A(.)|_T \nu_\infty |f|_T + |f|_T.$$

Thus

$$|x|_T \le \nu_\infty \left(\frac{\tilde{V}_d(R)(\nu_\infty \|A(.)\|_{C(0,\infty)} + 1)}{1-\psi_R} + 1 \right) |f|_T$$

$$= \nu_\infty (\frac{\psi_R}{1-\psi_R} + 1)|f|_T = \frac{\nu_\infty}{1-\psi_R}|f|_T.$$

Letting $T \to \infty$, we prove the assertion of the theorem. □

4.13 Determinants of Differential Delay Systems

Let C and C_1 be $n \times n$-matrices. Then as it is shown in Sect. 2.10, the spectral norm,

$$|det\ C - det\ C_1| \le n\|C - C_1\| M_2^{n-1}, \tag{13.1}$$

where $M_2 := \max\{\|C\|, \|C_1\|\}$. *Throughout this section* $\|C\|$ *means the spectral norm of matrix* C.

Now let

$$K(\lambda) = \lambda I - \int_0^\eta e^{\lambda s} dR_0(s)$$

and

$$K_1(\lambda) = \lambda I - \int_0^\eta e^{\lambda s} dR_1(s),$$

where $R_0(s)$ is the same as above and $R_1(s) = (r_{ij}^1(s))_{i,j=1}^n$ is a real $n \times n$-matrix-valued function defined on a finite segment $[0, \eta]$ whose entries have bounded variations and finite numbers of jumps.

From (13.1) it follows

$$|det\ K(\lambda) - det\ K_1(\lambda)| \le n\|K(\lambda) - K_1(\lambda)\|[\max\{\|K(\lambda)\|, \|K_1(\lambda)\|\}]^{n-1}.$$

But

$$\|K(i\omega)\| \le |\omega| + \int_0^\eta \|dR_0(s)\| = |\omega| + var(R_0)$$

and

$$\|K_1(i\omega)\| \le |\omega| + var(R_1) \quad (\omega \in \mathbf{R}).$$

In addition,

$$\|K(i\omega) - K_1(i\omega)\| \le \int_0^\eta \|d(R_0(s) - R_1(s))\| = var\ (R_1 - R_0).$$

Here according to the definition of $var(R_0)$, the expression $var\ (R_1 - R_0)$ means the spectral norm of the matrix $\left(var\ (r_{ij}^1 - r_{ij})\right)_{i,j=1}^n$. Put

$$\nu = \max\{var\ (R_0), var\ (R_1)\}.$$

Then we obtain

$$|det\ K(i\omega) - det\ K_1(i\omega)| \le \Delta(\omega) \ \ (\omega \in \mathbf{R}), \tag{13.2}$$

where

$$\Delta(\omega) := n\, var\ (R_1 - R_0)\ (|\omega| + \nu)^{n-1}.$$

Now Lemma 3.9 implies

Lemma 4.27 *Let $K(\lambda)$ be stable and $|det\ K(i\omega)| > \Delta(\omega)$ for all $\omega \in \mathbf{R}$. Then $K_1(\lambda)$ is also stable. Moreover,*

$$|det\ K_1(i\omega, t)| \ge |det\ K(i\omega)| - \Delta(\omega) \ \ (\omega \in \mathbf{R}).$$

We need also the following result.

Lemma 4.28 *One has*

$$\inf_{-\infty \le \omega \le \infty} |det\ \ K(i\omega)| = \inf_{-2\, var(R_0) \le \omega \le 2\, var(R_0)} |det\ K(i\omega)|.$$

Proof We have

$$\lambda_j(K(i\omega)) = i\omega - \lambda_j(\int_0^\eta e^{-i\omega} dR_0).$$

Since

$$|\lambda_j(\int_0^\eta e^{-i\omega} dR_0)| \le \|\int_0^\eta e^{-i\omega} dR_0\| \le var(R_0),$$

one has

$$|\lambda_j(K(i\omega))| \ge |\omega| - var(R_0) \ge var(R_0) \ \ (|\omega| \ge 2\ var(R_0)).$$

But

$$|\lambda_j(K(0))| = |\lambda_j(\int_0^\eta dR_0)| \le \|R_0(\eta) - R_0(0)\| \le var(R_0),$$

we can assert that the minimum of $|\lambda_j(K(i\omega))|$ is attained on $|\omega| \le 2\ var(R_0)$. Hence it follows that the minimum of $|det\ (K(i\omega))|$ is attained on $|\omega| \le 2\ var(R_0)$. As claimed. □

From the previous lemma we have

$$\Delta(\omega) \leq \Delta(2\nu) = n\ var\ (R_1 - R_0)(3\nu)^{n-1} \quad (\omega \leq 2\nu)$$

and

$$\begin{aligned}\inf_{-\infty\leq\omega\leq\infty} |det\ K_1(i\omega)| &= \inf_{-2\nu\leq\omega\leq 2\nu} |det\ K_1(i\omega)| \\ &\geq \inf_{-2\nu\leq\omega\leq 2\nu} |det\ K(i\omega)| - \Delta(2\nu) \\ &= \inf_{-2var(R_0)\leq\omega\leq 2var(R_0)} |det\ K(i\omega)| - \Delta(2\nu).\end{aligned}$$

Now Lemmas 4.27 and 4.28 imply

Theorem 4.6 *Let $K(\lambda)$ be stable and*

$$\inf_{-2\ var(R_0)\leq\omega\leq 2\ var(R_0)} |det\ K(i\omega)| > n\ var\ (R_1 - R_0)(3\nu)^{n-1}.$$

Then $K_1(\lambda)$ is also stable. Moreover,

$$\inf_{-\infty\leq\omega\leq\infty} |det\ K_1(i\omega)| > \inf_{-2\ var(R_0)\leq\omega\leq 2\ var(R_0)} |det\ K(i\omega)| - n\ var(R_1 - R_0)\ (3\nu)^{n-1}.$$

4.14 Comments

The material of this chapter is based on the book Gil' (2013a).

Chapter 5
Linear Autonomous NDEs

This chapter is devoted to vector linear autonomous NDEs. We derive estimates for the L^p- and C-norms of the characteristic matrix functions and fundamental solutions of the considered equations, as well as establish stability conditions for autonomous NDEs.

In this chapter $C(a, b) = C([a, b], \mathbf{C}^n)$, $C^1(a, b) = C^1([a, b], \mathbf{C}^n)$ and $L^p(a, b) = L^p([a, b], \mathbf{C}^n)$ $(p \ge 1)$.

5.1 Homogeneous Autonomous NDEs

For a positive constant $\eta < \infty$ consider the problem

$$\dot{y}(t) - \int_0^\eta d\tilde{R}_0(\tau)\dot{y}(t-\tau) = \int_0^\eta dR_0(\tau)y(t-\tau) \ (t > 0), \tag{1.1}$$

$$y(t) = \phi(t) \quad \text{for } -\eta \le t \le 0, \tag{1.2}$$

where $\dot{y}(t)$ is the right derivative of $y(t)$; $\phi \in C^1(-\eta, 0)$ is given;

$$R_0(s) = \left(r_{ij}(s)\right)_{i,j=1}^n \quad \text{and} \quad \tilde{R}_0(s) = \left(\tilde{r}_{ij}(s)\right)_{i,j=1}^n$$

are real $n \times n$-matrix-valued functions defined on $[0, \eta]$, whose entries have bounded variations

$$var(r_{ij}) \quad \text{and} \quad var(\tilde{r}_{ij}) \quad (i, j = 1, \dots, n),$$

respectively. In addition, *$\tilde{R}_0$ and R_0 have finite numbers of jumps and $\tilde{R}_0$ does not have a jump at zero.*

M.I. Gil', *Stability of Neutral Functional Differential Equations*,
Atlantis Studies in Differential Equations 3, DOI 10.2991/978-94-6239-091-1_5

The integrals in (1.1) are understood as the Lebesgue-Stieltjes integrals. Recall that $var(R_0) := \|(var(r_{ij}))_{i,j=1}^n\|_n$ (the spectral norm of matrix $(var(r_{ij}))_{i,j=1}^n$). Similarly $var(\tilde{R}_0)$ is defined. It is assumed that

$$var\,(\tilde{R}_0) < 1. \tag{1.3}$$

A solution of problem (1.1), (1.2) is an absolutely continuous vector valued function $y(t)$ defined on $[-\eta, \infty)$ and satisfying (1.1) and (1.2). Equation (1.1) is equivalent to the equation

$$y(t) - \int_0^\eta d\tilde{R}_0(\tau)y(t-\tau) = \phi(0) - \int_0^\eta d\tilde{R}_0(\tau)\phi(-\tau) + \int_0^t \int_0^\eta dR_0(\tau)y(t_1-\tau)dt_1 \tag{1.4}$$

for all $t > 0$. Again put

$$E_0 v(t) = \int_0^\eta dR_0(\tau)v(t-\tau)$$

and

$$\tilde{E}_0 v(t) = \int_0^\eta d\tilde{R}_0(\tau)v(t-\tau)$$

for a vector function $v(t) = (v_k(t))$.

Lemma 5.1 *Let condition* (1.3) *hold. Then a solution of Eq.* (1.4) *with condition* (1.2) *satisfies the a priory estimate*

$$\|y(t)\|_n \le \frac{c_0 + c_2 t}{1 - var(\tilde{R}_0)} \|\phi\|_{C(-\eta,0)}\, exp\left[\frac{t\; var(R_0)}{1 - var(\tilde{R}_0)}\right] \quad (t \ge 0), \tag{1.5}$$

where c_0, c_2 are positive constants independent of ϕ.

Proof For a vector function $v(t) = (v_k(t))$ again introduce the norm

$$\|v\|_{\hat{C}([a,b],\mathbf{C}^n)} = [\sum_{k=1}^n \sup_{a\le t\le b} |v_k(t)|^2]^{1/2}$$

(see Sect. 1.8). Recall that

$$\|v\|_{C([a,b],\mathbf{C}^n)} = \sup_{a\le t\le b} [\sum_{k=1}^n |v_k(t)|^2]^{1/2}.$$

Due to Lemma 1.8,

$$\|E_0 v\|_{\hat{C}([0,T],\mathbf{C}^n)} \leq var(R_0)\|v\|_{\hat{C}([-\eta,T],\mathbf{C}^n)} \quad (T > 0)$$

and therefore

$$\|\tilde{E}_0 v\|_{\hat{C}([0,T],\mathbf{C}^n)} \leq var(\tilde{R}_0)\|v\|_{\hat{C}([-\eta,T],\mathbf{C}^n)}.$$

Now (1.4) implies

$$\|y\|_{\hat{C}([0,t],\mathbf{C}^n)} \leq \|\phi(0)\|_n + var(\tilde{R}_0)\|\phi\|_{\hat{C}([-\eta,0],\mathbf{C}^n)}$$
$$+ var(\tilde{R}_0)\|y\|_{\hat{C}([-\eta,t],\mathbf{C}^n)} + var(R_0)\int_0^t \|y\|_{\hat{C}([-\eta,s],\mathbf{C}^n)} ds.$$

Denoting $m(t) = \|y\|_{\hat{C}([0,t],\mathbf{C}^n)}$ and $b = \|\phi\|_{\hat{C}([-\eta,0],\mathbf{C}^n)}$, we have

$$m(t) \leq b(1 + var(\tilde{R}_0)) + var(\tilde{R}_0)(b + m(t)) + var(R_0)\int_0^t (m(s) + b) ds.$$

Hence,

$$m(t) \leq (c_1 + t\, var(R_0))b + var(\tilde{R}_0)m(t) + var(R_0)\int_0^t m(s)ds,$$

where

$$c_1 = 1 + 2var(\tilde{R}_0).$$

Consequently,

$$m(t) \leq \frac{1}{1 - var(\tilde{R}_0)}[(c_1 + var(R_0)t)b + var(R_0)\int_0^t m(s)ds].$$

Now the Gronwall lemma yields

$$m(t) \leq \frac{c_1 + var(R)t}{1 - var(\tilde{R}_0)} \|\phi\|_{\hat{C}(-\eta,0)} exp\, [\frac{t\, var(R_0)}{1 - var(\tilde{R}_0)}]. \tag{1.6}$$

Since

$$\|y\|_{C([0,t),\mathbf{C}^n)} \leq \|y\|_{\hat{C}([0,t),\mathbf{C}^n)}$$

and

$$\|\phi\|^2_{\hat{C}([-\eta,0],\mathbf{C}^n)} \le n\|\phi\|^2_{C([-\eta,0],\mathbf{C}^n)}$$

(see Sect. 1.8), inequality (1.6) yields the required result. □

Inequality (1.5) enables us to apply the Laplace transform to problem (1.1), (1.2). Introduce the matrix-valued function

$$Q(z) = Iz - z\int_0^\eta exp(-zs)d\tilde{R}_0(s) - \int_0^\eta exp(-zs)dR_0(s) \ (z \in \mathbf{C}).$$

It is called *the characteristic function to Eq.* (1.1); the zeros of $det\ Q(\lambda)$ are called *the characteristic values of* $Q(.)$; $\lambda \in \mathbf{C}$ is *a regular value of* $Q(.)$, if $det\ Q(\lambda) \neq 0$.

Throughout this chapter it is assumed that all the characteristic values of $Q(.)$ *are in the open left half-plane* C_-.

Below we also suggest some conditions that provide the location of the characteristic values in C_-.

Lemma 5.2 *Let all the characteristic values of* $Q(.)$ *be in* C_- *and condition* (1.3) *hold. Then for any* $\phi \in C^1(-\eta, 0)$, *problem* (1.1), (1.2) *has a unique solution* y *which can be represented as*

$$y(t) = \frac{1}{2\pi}\int_{-\infty}^{\infty} e^{i\omega t}Q^{-1}(i\omega)\Psi(\phi, i\omega)d\omega$$

where

$$\Psi(\phi, z) = \phi(0) + \int_0^\eta d\tilde{R}_0(\tau)\left(\int_{-\tau}^0 e^{-z(t_1+\tau)}\dot{\phi}(t_1)dt_1 - e^{-z\tau}\phi(0)\right)$$
$$+\int_0^\eta dR_0(\tau)\int_{-\tau}^0 e^{-z(t_1+\tau)}\phi(t_1)dt_1.$$

Proof The existence and uniqueness of solutions to (1.1), (1.2) easily follow from the previous lemma and the compactness of the linear operator defined by the right-hand part of (1.4) in space $C(0, T)$ for $T < \infty$. Put

$$\hat{y}(z) = \int_0^\infty e^{-zt}y(t)dt,$$

where $y(t)$ is a solution of problems (1.1), (1.2). From the previous lemma it follows that the integrals converges absolutely al least in the half-plane $Rez > var(R_0)$. Moreover, from (1.1) and (1.3) it is also follows that $\dot{y}$ is also of exponential growth. In addition, we can write

$$\begin{aligned}\int_0^\infty e^{-zt}\int_0^\eta d\tilde{R}_0(\tau)\dot{y}(t-\tau)dt &= \int_0^\eta d\tilde{R}_0(\tau)\int_0^\infty e^{-zt}\dot{y}(t-\tau)dt\\ &= \int_0^\eta d\tilde{R}_0(\tau)\int_{-\tau}^\infty e^{-z(t_1+\tau)}\dot{y}(t_1)dt_1\\ &= \int_0^\eta d\tilde{R}_0(\tau)(ze^{-z\tau}\hat{y}(z) - e^{-z\tau}y(0)\\ &\quad + \int_{-\tau}^0 e^{-z(t_1+\tau)}\dot{\phi}(t_1)dt_1)\end{aligned}$$

and

$$\begin{aligned}\int_0^\infty e^{-zt}\int_0^\eta dR_0(\tau)y(t-\tau)dt &= \int_0^\eta dR_0(\tau)\int_0^\infty e^{-zt}y(t-\tau)dt\\ &= \int_0^\eta dR_0(\tau)\int_{-\tau}^\infty e^{-z(t_1+\tau)}y(t_1)dt_1\\ &= \int_0^\eta dR_0(\tau)(e^{-z\tau}\hat{y}(z) + \int_{-\tau}^0 e^{-z(t_1+\tau)}\phi(t_1)dt_1).\end{aligned}$$

Interchanging the Stieltjes integration with the improper Riemann integration is justified by the Fubini theorem.

Taking the Laplace transforms of the both sides of Eq. (1.1), we get

$$\begin{aligned}&z\hat{y}(z) - \phi(0) - \int_0^\eta d\tilde{R}_0(\tau)(ze^{-z\tau}\hat{y}(z) - e^{-z\tau}\phi(0) + \int_{-\tau}^0 e^{-z(t_1+\tau)}\dot{\phi}(t_1)dt_1)\\ &= \int_0^\eta dR_0(\tau)(e^{-z\tau}\hat{y}(z) + \int_{-\tau}^0 e^{-z(t_1+\tau)}\phi(t_1)dt_1).\end{aligned}$$

Or $Q(z)\hat{y}(z) = \Psi(\phi, z)$. Hence applying the inverse Laplace transform and taking into account that all the characteristic values of $Q(.)$ are in the open left half-plane C_- we obtain the required result. □

5.2 The Fundamental Solutions to NDEs

An absolutely continuous matrix valued function $Z(t)$ satisfying the equation

$$\dot{Z}(t) - \int_0^\eta d\tilde{R}_0(\tau)\dot{Z}(t-\tau) = \int_0^\eta dR_0(\tau)Z(t-\tau)\ (t > 0) \tag{2.1}$$

and the conditions

$$Z(t) = 0 \quad (t < 0), \qquad Z(0) = I \tag{2.2}$$

will be called *the fundamental solution to Eq.* (1.1). Recall that I is the unit matrix.

Lemma 5.3 *Let all the characteristic values of $Q(.)$ be in C_- and condition* (1.3) *hold. Then*

$$Z(t) = \frac{1}{2\pi}\int_{-\infty}^{\infty} e^{it\omega}Q^{-1}(i\omega)d\omega \quad (t \geq 0) \tag{2.3}$$

and the integral exists in the sense of the main value.

Proof Repeating the arguments of the proof of Lemma 5.1, it can be shown that $Z(t)$ is of exponential growth, and therefore (2.1) admits the Laplace transform. So denoting by $Y(z)$ the Laplace transform to $Z(t)$, from (2.1) we have

$$zY(z) - Y(0) - \int_0^\eta e^{-z\tau}d\tilde{R}_0(\tau)\int_0^\infty e^{-zt}\dot{Z}(t-\tau)dt = \int_0^\eta e^{-\tau z}dR_0(\tau)Y(z).$$

Integrating by parts, we have

$$\int_0^\infty e^{-zt}\dot{Z}(t-\tau)dt = -Z(-\tau) + z\int_\tau^\infty e^{-zt}Z(t-\tau)dt = -Z(-\tau) + e^{-z\tau}zY(z).$$

But

$$\int_0^\eta e^{-z\tau} d\tilde{R}_0(\tau)Z(-\tau) = \int_0^\varepsilon e^{-z\tau} d\tilde{R}_0(\tau)Z(-\tau) + \int_\varepsilon^\eta e^{-z\tau} d\tilde{R}_0(\tau)Z(-\tau)$$

$$= \int_0^\varepsilon e^{-z\tau} d\tilde{R}_0(\tau)Z(-\tau) \to 0 \quad \text{as } \varepsilon \to 0,$$

since $\tilde{R}_0(\tau)$ does not have a jump at zero. Hence

$$zY(z) - I - z\int_0^\eta e^{-z\tau} d\tilde{R}_0(\tau)Y(z) = \int_0^\eta e^{-\tau z} dR_0(\tau)Y(z).$$

and therefore

$$Y(z) = Q^{-1}(z).$$

Due to the inverse Laplace transform we obtain equality (2.3). □

Lemma 5.4 *A solution of problem* (1.1), (1.2) *admits the representation*

$$y(t) = Z(t)\phi(0) + \int_0^\eta \left(\int_{-\tau}^0 (Z(t - t_1 - \tau) d\tilde{R}_0(\tau)\dot{\phi}(t_1)dt_1 - Z(t - \tau)d\tilde{R}_0(\tau)\phi(0) \right)$$

$$+ \int_0^\eta \int_{-\tau}^0 Z(t - t_1 - \tau)dR_0(\tau)\phi(t_1)dt_1 \quad (t \geq 0).$$

Proof From Lemma 5.2 it follows that

$$y(t) = M_1(t) + M_2(t) + M_3(t),$$

where

$$M_1(t) = \frac{1}{2\pi i} \int_{-i\infty}^{i\infty} e^{zt} Q^{-1}(z)\, dz\, \phi(0) = Z(t)\phi(0),$$

$$M_2(t) = \frac{1}{2\pi i} \int_{-i\infty}^{i\infty} e^{zt} Q^{-1}(z) \int_0^\eta d\tilde{R}_0(\tau) \left(\int_{-\tau}^0 e^{-z(t_1+\tau)}\dot{\phi}(t_1)dt_1 - e^{-z\tau}\phi(0) \right) dz$$

and

$$M_3(t) = \frac{1}{2\pi i} \int_{-i\infty}^{i\infty} e^{zt} Q^{-1}(z) \int_0^{\eta} dR_0(\tau) \int_{-\tau}^{0} e^{-z(t_1+\tau)} \phi(t_1) dt_1 dz.$$

But

$$\begin{aligned}
&\frac{1}{2\pi i} \int_{-i\infty}^{i\infty} e^{zt} Q^{-1}(z) \int_0^{\eta} d\tilde{R}_0(\tau) \int_{-\tau}^{0} e^{-z(t_1+\tau)} \dot{\phi}(t_1) dt_1 \\
&= \int_0^{\eta} \int_{-\tau}^{0} (\frac{1}{2\pi i} \int_{-i\infty}^{i\infty} e^{z(t-t_1-\tau)} Q^{-1}(z) dz) d\tilde{R}_0(\tau) \dot{\phi}(t_1) dt_1 \\
&= \int_0^{\eta} \int_{-\tau}^{0} Z(t - t_1 - \tau) d\tilde{R}_0(\tau) \dot{\phi}(t_1) dt_1.
\end{aligned}$$

In addition,

$$\frac{1}{2\pi i} \int_{-i\infty}^{i\infty} e^{zt} Q^{-1}(z) \int_0^{\eta} d\tilde{R}_0(\tau) e^{-z\tau} \phi(0) dz = \int_0^{\eta} Z(t - \tau) d\tilde{R}_0(\tau) \phi(0).$$

So

$$M_2(t) = \int_0^{\eta} \left(\int_{-\tau}^{0} Z(t - t_1 - \tau) d\tilde{R}_0(\tau) \dot{\phi}(t_1) dt_1 - Z(t - \tau) d\tilde{R}_0(\tau) \phi(0) \right).$$

Similarly,

$$M_3(t) = \int_0^{\eta} \int_{-\tau}^{0} Z(t - t_1 - \tau) dR_0(\tau) \phi(t_1) dt_1.$$

This proves the required result. □

5.3 Nonhomogeneous Autonomous NDEs

Consider the non-homogeneous equation

$$\dot{x}(t) - \int_0^{\eta} d\tilde{R}_0(\tau) \dot{x}(t - \tau) = \int_0^{\eta} dR_0(\tau) x(t - \tau) + f(t) \quad (t > 0), \tag{3.1}$$

$$x(t) = 0 \quad \text{for } -\eta \le t \le 0, \tag{3.2}$$

with a function f bounded and integrable on each finite segment $[0, T]$.

A solution of problem (3.1), (3.2) is an absolutely continuous vector valued function $x(t)$ defined on $[-\eta, \infty)$ and satisfying (3.1), (3.2). (3.1) is equivalent to the equation

$$x(t) - \int_0^\eta d\tilde{R}_0(\tau)x(t-\tau) = \int_0^t f(t)dt + \int_0^t \int_0^\eta dR_0(\tau)x(t_1-\tau)dt_1 \quad (t \ge 0). \tag{3.3}$$

Lemma 5.5 *Let condition* (1.3) *hold. Then a solution of Eq.* (3.3) *satisfies the apriory estimate*

$$\|x(t)\|_n \le \frac{c_1 \|f\|_{L^1(0,t)}}{1 - var(\tilde{R}_0)} exp\, [\frac{t\ var(R_0)}{1 - var(\tilde{R}_0)}] \quad (t \ge 0) \tag{3.4}$$

where c_1 is a constant independent of f.

Proof For a vector function $v(t) = (v_k(t))$ again use the norm

$$\|v\|_{\hat{C}([a,b],\mathbf{C}^n)} = [\sum_{k=1}^n \sup_{a\le t\le b} |v_k(t)|^2]^{1/2}$$

(see Sect. 1.8). Recall that it is equivalent to the norm

$$\|v\|_{C([a,b],\mathbf{C}^n)} = \sup_{a\le t\le b} [\sum_{k=1}^n |v_k(t)|^2]^{1/2}$$

for all $-\infty < a < b \le \infty$. Due to Lemma 1.8,

$$\|E_0 v\|_{\hat{C}([0,T],\mathbf{C}^n)} \le var(R_0)\|v\|_{\hat{C}([-\eta,T],\mathbf{C}^n)},\ \|\tilde{E}_0 v\|_{\hat{C}([0,T],\mathbf{C}^n)} \le var(\tilde{R}_0)\|v\|_{\hat{C}([-\eta,T],\mathbf{C}^n)}.$$

Now (3.3) implies

$$\|x\|_{\hat{C}([0,T],\mathbf{C}^n)} \le const\|f\|_{L^1(0,T)} + var(\tilde{R}_0)\|x\|_{\hat{C}([0,T],\mathbf{C}^n)} + var(R_0)\int_0^t \|x\|_{\hat{C}([0,s],\mathbf{C}^n)}ds.$$

Put $m(t) = \|x\|_{\hat{C}([0,t],\mathbf{C}^n)}$. Then we can write

$$m(t) \le const\|f\|_{L^1(0,T)} + var(\tilde{R}_0)m(t) + var(R_0)\int_0^t m(s)ds.$$

Hence,

$$m(t) \le \frac{1}{1 - var(\tilde{R}_0)}[const\ \|f\|_{L^1(0,t)} + var(R_0)\int_0^t m(s)ds].$$

Now the Gronwall lemma yields the inequality

$$m(t) \le \frac{const\ \|f\|_{L^1(0,t)}}{1 - var(\tilde{R}_0)} exp\,[\frac{t\ var(R_0)}{1 - var(\tilde{R}_0)}]\quad (t \ge 0).$$

Since $\|x\|_{C([0,t],\mathbf{C}^n)} \le \|y\|_{\hat{C}([0,t],\mathbf{C}^n)}$ this yields the required inequality. □

From the previous lemma we have the existence and uniqueness of solutions to problem (3.1), (3.2). Inequality (3.4) enables us to apply the Laplace transform to Eq. (3.1) with $f \in L^2(0,\infty) \cap L^1(0,\infty)$. It gives us $Q(z)\hat{x}(z) = \hat{f}(z)$ where $\hat{x}(z)$ and $\hat{f}(z)$ are the Laplace transforms of $x(t)$ and $f(t)$, respectively. Consequently,

$$\hat{x}(z) = Q^{-1}(z)\hat{f}(z) \tag{3.5}$$

for any regular z. Hence, by the inverse Laplace transform we obtain

$$x(t) = \int_0^t Z(t-s)f(s)ds. \tag{3.6}$$

This formula is called the Variation of Constants formula to Eq. (3.1). Moreover, from (3.5) it follows

$$\|\hat{x}(i\omega)\|^2_{L^2(-\infty,\infty)} := \int_{-\infty}^{\infty} \|\hat{x}(i\omega)\|_n^2 d\omega$$
$$\le \sup_{-\infty\le\omega\le\infty} \|Q^{-1}(i\omega)\|_n^2 \|\hat{f}(i\omega)\|^2_{L^2(-\infty,\infty)}.$$

Now the Parseval equality implies

$$\|x\|_{L^2(0,\infty)} \le \sup_{-\infty\le\omega\le\infty} \|Q^{-1}(i\omega)\|_n \|f\|_{L^2(0,\infty)}. \tag{3.7}$$

The following quantities play an essential role hereafter:

$$\zeta_0 := \frac{2\, var(R_0)}{1 - var(\tilde{R}_0)},$$

and

$$\theta(Q) := \sup_{-\zeta_0 \le \omega \le \zeta_0} \|Q^{-1}(i\omega)\|_n.$$

We need the following lemma

Lemma 5.6 *Let all the characteristic values of* $Q(.)$ *be in the open left half-plane* C_- *and condition* (1.3) *hold. Then the equality*

$$\sup_{-\infty \le \omega \le \infty} \|Q^{-1}(i\omega)\|_n = \theta(Q)$$

is valid.

Proof Without loss of generality assume that $R_0(0-) = 0$. We have $-Q(0) = R_0(\eta)$ and

$$\|Q^{-1}(0)\|_n = \|R_0^{-1}(\eta)\|_n \ge \frac{1}{\|R_0(\eta)\|_n} \ge \frac{1}{var(R_0)}.$$

In addition,

$$\begin{aligned}\|Q(i\omega)u\|_n &\ge (|\omega|(1 - var(\tilde{R}_0)) - var(R_0))\|u\|_n \\ &\ge var(R_0)\|u\|_n \quad (\omega \in \mathbf{R}, |\omega| \ge \zeta_0;\ u \in \mathbf{C}^n).\end{aligned}$$

So

$$\|Q^{-1}(i\omega)\|_n \le \frac{1}{var(R_0)} \le \|Q^{-1}(0)\|_n \quad (|\omega| \ge \zeta_0).$$

Thus the maximum of $\|Q^{-1}(i\omega)\|_n$ is attained on $[-\zeta_0, \zeta_0]$. As claimed. □

Making use (3.7) and the previous lemma, we get

$$\|x\|_{L^2(0,\infty)} \le \theta(Q)\|f\|_{L^2(0,\infty)}. \tag{3.8}$$

If for a given function f a problem (3.1), (3.2) has a solution x, then as above, the operator $\hat{Z}$ defined by $x = \hat{Z}f$ is called the *Cauchy operator* to Eqs. (1.1) and (3.1). Due to the Variation of Constants formula

$$\hat{Z}f(t) = \int_0^t Z(t-s)f(s)ds.$$

Now (3.8) implies

Theorem 5.1 Let condition (1.3) hold and all the zeros of $det\ Q(z)$ be in C_-. Then the Cauchy operator to (3.1) satisfies the inequality

$$\|\hat{Z}\|_{L^2(0,\infty)} \le \theta(Q).$$

5.4 Estimates for $\theta(Q)$

Recall that $N_2(A)$ is the Hilbert-Schmidt (Frobenius) norm and

$$g(A) = (N_2^2(A) - \sum_{k=1}^{n} |\lambda_k(A)|^2)^{1/2}$$

(see Sect. 2.1); besides

$$g^2(A) \le N_2^2(A - A^*)/2 \quad \text{and} \quad g(e^{i\tau}A + zI) = g(A) \tag{4.1}$$

for all $\tau \in \mathbf{R}$ and $z \in \mathbf{C}$. About other properties of $g(A)$ see Sect. 2.1.

Due to Corollary 2.1, for an invertible $n \times n$-matrix A one has

$$\|A^{-1}\|_n \le \sum_{k=0}^{n-1} \frac{g^k(A)}{\sqrt{k!}\rho^{k+1}(A)}, \tag{4.2}$$

where $\rho(A) = \min_{k=1,\dots,n} |\lambda_k(A)|$.

Put

$$B(z) = z\int_0^{\eta} exp(-zs)dR_0(s) + \int_0^{\eta} exp(-zs)dR_0(s).$$

So $Q(z) = zI - B(z)$. By (4.1) $g(B(z)) = g(Q(z))$. Thanks to (4.2), for any regular value z of $Q(.)$, the inequality

$$\|[Q(z)]^{-1}\|_n \le \Gamma(Q(z))\ (z \in \mathbf{C}) \tag{4.3}$$

is valid, where

$$\Gamma(Q(z)) = \sum_{k=0}^{n-1} \frac{g^k(B(z))}{\sqrt{k!}\rho^{k+1}(Q(z))}$$

and $\rho(Q(z))$ is the smallest absolute value of the eigenvalues of $Q(z)$:

$$\rho(Q(z)) = \min_{k=1,\dots,n} |\lambda_k(Q(z))|.$$

If $B(z)$ is a normal matrix, then $g(B(z)) = 0$, and

$$\|[Q(z)]^{-1}\|_n \leq \frac{1}{\rho(Q(z))}.$$

For example, that relation holds, if

$$Q(z) = zI - \tilde{A}ze^{-zh_1} - Ae^{-zh_2},$$

where A and $\tilde{A}$ are commuting Hermitian matrices, h_1, h_2 are nonnegative constants.

Due to (4.3) we arrive at

Lemma 5.7 *One has*

$$\theta(Q) \leq \Gamma_0(Q), \quad \textit{where } \Gamma_0(Q) := \sup_{-\zeta_0 \leq \omega \leq \zeta_0} \Gamma(Q(i\omega)).$$

Furthermore, Theorem 2.2 implies

$$\|A^{-1}\, det\ (A)\|_n \leq \frac{N_2^{n-1}(A)}{(n-1)^{(n-1)/2}}$$

for any invertible $n \times n$-matrix A. Hence, for any regular point z of $Q(.)$, one has

$$\|Q^{-1}(z)\|_n \leq \frac{N_2^{n-1}(Q(z))}{(n-1)^{(n-1)/2}|det\ (Q(z))|},$$

and thus we obtain the following result.

Lemma 5.8 *Under condition* (1.3), *the inequality*

$$\theta(Q) \leq \theta_d(Q), \quad \textit{where } \theta_d(Q) := \sup_{-\zeta_0 \leq \omega \leq \zeta_0} \frac{N_2^{n-1}(Q(i\omega))}{(n-1)^{(n-1)/2}|det\ (Q(i\omega))|}$$

is true.

5.5 Integrals of Characteristic Matrix Functions to NDEs

Put

$$\|Q^{-1}(i\omega)\|_{L^2(-\infty,\infty)} := \left(\int_{-\infty}^{\infty} \|Q^{-1}(i\omega)\|_n^2 d\omega\right)^{1/2}.$$

Lemma 5.9 *Let condition* (1.3) *hold. Then the inequality*

$$\|Q^{-1}(i\omega)\|^2_{L^2(-\infty,\infty)} \le \frac{2\theta^2(Q)var(R_0)+4\theta(Q)}{1-var(\tilde R_0)}$$

is valid.

Proof We have

$$\|Q(i\omega)u\|_n = \|i\omega(I-\int_0^\eta e^{-i\omega\tau}d\tilde R_0(\tau))u - \int_0^\eta e^{-i\omega\tau}dR_0(\tau)u\|_n$$
$$\ge (|\omega|(1-\int_0^\eta \|d\tilde R_0(\tau)\|_n) - \int_0^\eta \|dR_0(\tau)\|_n)\|u\|_n$$
$$= (|\omega|(1-var\,(\tilde R_0)) - var\,(R_0))\|u\|_n \quad (u\in \mathbf{C}^n),$$

provided

$$\omega\in\mathbf{R} \text{ with } |\omega| > \frac{var(R_0)}{1-var(\tilde R_0)}. \tag{5.1}$$

Therefore, under condition (5.1) we have

$$\|Q^{-1}(i\omega)\|_n \le \frac{1}{|\omega|(1-var\,(\tilde R_0)) - var\,(R_0)} \tag{5.2}$$

For the brevity, in this proof put

$$m(\omega)=\|Q^{-1}(i\omega)\|_n \quad (\omega\in\mathbf{R}),\, c=\frac{1}{1-var(\tilde R_0)},\, v_0=\frac{var(R_0)}{1-var(\tilde R_0)}=\zeta_0/2$$

and $\theta=\theta(Q)$. Consequently, $v_0 = c\,var(R_0)$, and

$$m(\omega)\le \frac{1}{|\omega|(1-var\,(\tilde R_0)) - var\,(R_0)} \quad (|\omega|>v_0).$$

We should estimate the integral

$$J=\int_{-\infty}^{\infty} m^2(s)ds.$$

For an $a>v_0$ we can write $J=J_1+J_2$, where

$$J_1=\int_{-a}^{a} m^2(s)ds,\ J_2=\int_{-\infty}^{-a} m^2(s)ds+\int_a^\infty m^2(s)ds.$$

Due to Lemma 5.6

$$\|m(\omega)\|_n \le \theta \quad \text{for all } \omega \in \mathbf{R}. \tag{5.3}$$

Thus

$$J_1 \le 2a\theta^2$$

and, in addition,

$$J_2 \le 2\int_a^\infty \frac{ds}{(s(1 - var\,(\tilde{R}_0)) - var\,(R_0))^2} = 2c^2 \int_a^\infty \frac{ds}{(s - v_0)^2} = 2c^2 \frac{1}{a - v_0}.$$

Hence $J \le 2f(a)$ where

$$f(a) = \theta^2 a + c^2 \frac{1}{a - v_0}.$$

Obviously,

$$df(a)/da = \theta^2 - \frac{c^2}{(a - v_0)^2}.$$

So the minimum of $f(a)$ is attained for

$$a = \frac{c}{\theta} + v_0$$

and therefore one can write

$$\frac{1}{2}J \le \theta^2(v_0 + \frac{c}{\theta}) + c^2\frac{\theta}{c} = \theta^2 v_0 + 2c\theta,$$

as claimed. □

Since,

$$\|Q(0)\|_n = \|\int_0^\eta dR_0(s)\|_n \le var(R_0)$$

we have

$$\theta(Q) \ge \|Q^{-1}(0)\|_n \ge \frac{1}{var(R_0)}.$$

Or

$$\theta(Q)var(R_0) \geq 1. \tag{5.4}$$

Now the previous lemma implies

$$\|Q^{-1}(i\omega)\|^2_{L^2(-\infty,\infty)} \leq \frac{6\, var(R_0)\theta^2(Q)}{1 - var(\tilde{R}_0)} = 3\theta^2(Q)\zeta_0. \tag{5.5}$$

Furthermore, put

$$\left\| \frac{dQ^{-1}(i\omega)}{d\omega} \right\|_{L^2(-\infty,\infty)} := \left(\int_{-\infty}^{\infty} \left\| \frac{dQ^{-1}(i\omega)}{d\omega} \right\|_n^2 \right)^{1/2}.$$

Recall that $vd(R_0)$ means the spectral norm of the matrix $(vd(r_{jk}))^n_{j,k=1}$, whose entries are

$$vd(r_{ij}) = \int_0^{\eta} s|dr_{jk}(s)|$$

(see Sect. 4.4). Recall also that

$$\zeta_0 = \frac{2var(R_0)}{1 - var(\tilde{R}_0)}$$

and denote

$$D(s) = 1 + var(\tilde{R}_0) + vd(R_0) + s\; vd(\tilde{R}_0) \quad (s \geq 0).$$

We need the following lemma.

Lemma 5.10 *Let condition* (1.3) *hold. Then the inequality*

$$\left\| \frac{dQ^{-1}(i\omega)}{d\omega} \right\|_{L^2(-\infty,\infty)} \leq \theta^2(Q)D(\zeta_0)\sqrt{3\zeta_0}$$

is true.

Proof We have

$$\frac{dQ^{-1}(z)}{dz} = -Q^{-1}(z)\frac{dQ(z)}{dz}Q^{-1}(z),$$

and consequently,

$$\left\| \frac{dQ^{-1}(i\omega)}{d\omega} \right\|_{L^2(-\infty,\infty)} \le \sup_{\omega \in \mathbf{R}} \left\| Q^{-1}(i\omega) \frac{dQ(i\omega)}{d\omega} \right\|_n \|Q^{-1}(i\omega)\|_{L^2(-\infty,\infty)}. \quad (5.6)$$

In addition,

$$\frac{dQ(i\omega)}{d\omega} = iI - i \int_0^\eta e^{-i\tau\omega} d\tilde{R}_0(\tau) - \omega \int_0^\eta \tau e^{-i\tau\omega} d\tilde{R}_0(\tau) + i \int_0^\eta \tau e^{-i\tau\omega} dR_0(\tau).$$

Simple calculations show that

$$\| \int_0^\eta \tau e^{-i\tau\omega} dR_0(\tau) \|_n \le \int_0^\eta \tau \|dR_0(\tau)\|_n \le vd(R_0)$$

and therefore,

$$\left\| \frac{dQ(i\omega)}{d\omega} \right\|_n \le D(|\omega|) \quad (\omega \in \mathbf{R}).$$

For the brevity put

$$v(\omega) := \|Q^{-1}(i\omega)\|_n \left\| \frac{dQ(i\omega)}{d\omega} \right\|_n .$$

Due (5.1) and (5.2) we have

$$v(s) \le f_1(s) \ (s > \zeta_0),$$

where

$$f_1(s) = \frac{D(s)}{s(1 - var\,(\tilde{R}_0)) - var\,(R_0)} = \frac{D(s)}{s} \cdot \frac{1}{1 - var\,(\tilde{R}_0) - \frac{var\,(R_0)}{s}}.$$

The function

$$\frac{D(s)}{s} = \frac{1}{s}(1 + var(\tilde{R}_0)) + vd(R_0)) + vd(\tilde{R}_0)$$

decreases and the function

$$1 - var\,(\tilde{R}_0) - \frac{var\,(R_0)}{s}$$

increases for $s \geq \zeta_0$. Thus

$$\nu(\omega) \leq f_1(\zeta_0) = \frac{D(\zeta_0)}{var\ (R_0)}$$

for $|\omega| > \zeta_0$. But $\|Q^{-1}(i\omega)\|_n \leq \theta(Q)$ and $D(|\omega|) \leq D(\zeta_0)$ for $|\omega| \leq \zeta_0$. Hence taking into account (5.4) we obtain $1/var\ (R_0) \leq \theta(K)$ and therefore

$$\nu(\omega) \leq \theta(Q)D(\zeta_0). \tag{5.7}$$

for all real ω. Now (5.6) implies

$$\left\|\frac{dQ^{-1}(i\omega)}{d\omega}\right\|_{L^2(-\infty,\infty)} \leq \|Q^{-1}(i\omega)\|_{L^2(-\infty,\infty)}\theta(Q)D(\zeta_0).$$

This and (5.5) yield

$$\left\|\frac{dQ^{-1}(i\omega)}{d\omega}\right\|^2_{L^2(-\infty,\infty)} \leq 3\ D^2(\zeta_0)\theta^4(Q)\zeta_0.$$

We thus have prove the required result. □

5.6 Integrals of Fundamental Solutions

By the Parseval equality we can write

$$\int_0^\infty \|Z(t)u\|_n^2 dt = \frac{1}{2\pi}\int_{-\infty}^\infty \|Q^{-1}(i\omega)u\|_n^2 d\omega \quad (u \in \mathbf{C}^n) \tag{6.1}$$

and

$$\int_0^\infty N_2^2(Z(t))dt = \frac{1}{2\pi}\int_{-\infty}^\infty N_2^2(Q^{-1}(i\omega))d\omega \leq n\int_{-\infty}^\infty \|Q^{-1}(i\omega)\|_n^2 d\omega.$$

Put

$$W(Q) = \left[\frac{n(\theta^2(Q)var(R_0) + 2\theta(Q))}{\pi(1 - var(\tilde{R}_0))}\right]^{1/2}.$$

Now Lemma 5.9 implies

Corollary 5.1 *Let condition* (1.3) *hold and all the characteristic values of* $Q(.)$ *be in* C_-. *Then*

$$\int_0^\infty N_2^2(Z(t))dt \le W^2(Q)$$

and

$$\int_0^\infty \|Z(t)u\|_n^2 dt \le \frac{1}{n} W^2(Q)$$

for all $u \in \mathbf{C}^n$ *with* $\|u\|_n = 1$.

According to (5.4) we can write

$$W^2(Q) \le \frac{3n\theta^2(Q)\zeta_0}{2\pi}.$$

Hence, the latter corollary implies

$$\int_0^\infty N_2^2(Z(t))dt \le \frac{3n\theta^2(Q)\zeta_0}{2\pi}. \tag{6.2}$$

Furthermore, the Laplace transform to $tZ(t)$ is $-dQ^{-1}(z)/dz$. So by the Parseval equality

$$\int_0^\infty t^2 N_2^2(Z(t))dt = \frac{1}{2\pi}\int_{-\infty}^\infty N_2^2(dQ^{-1}(is)/ds)ds \le \frac{n}{2\pi}\left\|\frac{dQ^{-1}(i\omega)}{d\omega}\right\|^2_{L^2(-\infty,\infty)}.$$

Now Lemma 5.10 yields

$$\int_0^\infty t^2 \|Z(t)\|_n^2 dt \le \frac{3n}{2\pi}\theta^4(Q)D^2(\zeta_0)\zeta_0. \tag{6.3}$$

Moreover, due to the Karlson inequality cf. (Mitrinovic, Chap. VIII) (see also Sect. 4.4.5), we can write

$$\|Z\|^4_{L^1(0,\infty)} \le \pi^2 \|Z\|^2_{L^2(0,\infty)} \|tZ\|^2_{L^2(0,\infty)}.$$

Now, applying inequalities (6.2) and (6.3), we arrive at our next result.

Corollary 5.2 *Let condition* (1.3) *hold and all the characteristic values of $Q(.)$ be in C_-. Then the fundamental solution to Eq.* (1.1) *is subject to the estimate*

$$\|Z\|^2_{L^1(0,\infty)} \le \frac{3n}{2}\theta^3(Q)D(\zeta_0)\zeta_0.$$

5.7 Norms of the Cauchy Operator

Taking into account Theorem 1.5 on the properties of convolutions, due to Corollary 5.2 we obtain the following result.

Theorem 5.2 Let condition (1.3) hold and all the characteristic values of $Q(.)$ be in C_-. Then the Cauchy operator to the neutral type Eq. (1.1) satisfies the inequality

$$\|\hat{Z}\|_{L^p(0,\infty)} \le \theta(Q)\sqrt{\frac{3}{2}\theta(Q)D(\zeta_0)\zeta_0} \quad (p \ge 1).$$

Besides, $\|\hat{Z}\|_{C(0,\infty)} = \|\hat{Z}\|_{L^\infty(0,\infty)}$.

Recall that some estimates for $\theta(Q)$ are given in Sect. 5.4. Additional estimates for $\theta(Q)$ are suggested below.

Equation (1.1) is said to be exponentially stable, if there are positive constants $\varepsilon > 0$ and m_ε independent of ϕ, such that the inequality

$$\|y(t)\|_n \le m_\varepsilon e^{-\varepsilon t}\|\phi\|_{C^1(-\eta,0)} \quad (t > 0)$$

holds for any solution $y(t)$ of problem (1.1), (1.2).

Now we are in a position to formulate our next result.

Corollary 5.3 *Let all the characteristic values of the neutral type Eq.* (1.1) *lie in C_- and condition* (1.3) *hold. Then Eq.* (1.1) *is exponentially stable.*

Proof Indeed, due to Lemma 5.4 a solution of problem (1.1), (1.2) satisfies the inequality

$$\|y\|_{L^2(0,\infty)} \le const\,\|\phi\|_{C^1(-\eta,0)}\|Z\|_{L^2(0,\infty)} = c_1\|\phi\|_{C^1(-\eta,0)} \; (c_1 = const).$$

From (1.1) it follows that

$$\begin{aligned}\|\dot{y}\|_{L^2(0,\infty)} &\le var(\tilde{R}_0)\|\dot{y}\|_{L^2(-\eta,\infty)} + var(R_0)\|y\|_{L^2(-\eta,\infty)} \\ &\le var(\tilde{R}_0)\|\dot{y}\|_{L^2(-\eta,\infty)} + var(R_0)\|y\|_{L^2(-\eta,\infty)} \\ &\le var(\tilde{R}_0)(\|\dot{y}\|_{L^2(-0,\infty)} + \|\dot{\phi}\|_{L^2(-\eta,0)}) + var(R_0)\|y\|_{L^2(-\eta,\infty)}.\end{aligned}$$

Hence,

$$\|\dot{y}\|_{L^2(0,\infty)} \leq \frac{1}{1 - var(\tilde{R}_0)}[var(\tilde{R}_0)\|\dot{\phi}\|_{L^2(-\eta,0)} + var(R_0)\|y\|_{L^2(-\eta,\infty)}]$$
$$\leq c_2\|\phi\|_{C^1(-\eta,0)} \quad (c_2 = const).$$

Making use 4.5 we have

$$\|y\|^2_{C(0,\infty)} \leq 2\|\dot{y}\|_{L^2(0,\infty)}\|y\|_{L^2(0,\infty)} \leq c_2 c_1 \|\phi\|^2_{C^1(-\eta,0)}. \tag{7.1}$$

Furthermore, substituting the equality

$$y(t) = y_\varepsilon(t)e^{-\varepsilon t}, \tag{7.2}$$

into (1.1), we obtain the equation

$$\dot{y}_\varepsilon(t) - \varepsilon y_\varepsilon(t) - \int_0^\eta e^{\varepsilon\tau} d\tilde{R}_0(\tau)\dot{y}_\varepsilon(t-\tau) + \varepsilon \int_0^\eta e^{\varepsilon\tau} d\tilde{R}_0(\tau) y_\varepsilon(t-\tau)$$
$$= \int_0^\eta e^{\varepsilon\tau} dR_0(\tau) y_\varepsilon(t-\tau). \tag{7.3}$$

Take $\varepsilon > 0$ sufficiently small in such a way that all the characteristic values of (7.3) remain in C_- and the inequality

$$e^{\varepsilon\eta} var(\tilde{R}_0) < 1 \tag{7.4}$$

holds. Under these conditions according to (7.1) a solution y_ε to Eq. (7.3) is uniformly bounded on the positive half-line. Hence (7.2) yields the exponential stability. □

5.8 Lower Estimates for Quasi-Polynomials

In this section we suggest estimates for quasi-polynomials which will be used in the present chapter below. Consider the function

$$k(z) = z\left(1 - \int_0^\eta e^{-\tau z} d\tilde{\mu}(\tau)\right) + \int_0^\eta e^{-\tau z} d\mu(\tau) \quad (z \in \mathbf{C}), \tag{8.1}$$

where $\mu = \mu(\tau)$ and $\tilde{\mu} = \tilde{\mu}(\tau)$ are nondecreasing functions defined on $[0, \eta]$, and satisfying the conditions

$$0 < var(\tilde{\mu}) < 1 \quad \text{and} \quad var(\mu) < \infty. \tag{8.2}$$

Put

$$v_1 = \frac{2\, var(\mu)}{1 - var(\tilde{\mu})}.$$

The following two lemmas are proved in Sect. 8.6 below.

Lemma 5.11 *The equality*

$$\inf_{-\infty \le \omega \le \infty} |k(i\omega)| = \inf_{-v_1 \le \omega \le v_1} |k(i\omega)|$$

is valid.

Lemma 5.12 *Let the conditions* (8.2),

$$\eta v_1 < \pi/2 \tag{8.3}$$

and

$$d_0 := \int_0^\eta \cos(v_1\tau) d\mu(\tau) - v_1 \int_0^\eta \sin(v_1\tau) d\tilde{\mu}(\tau) > 0 \tag{8.4}$$

hold. Then all the zeros of $k(.)$ *are in* C_- *and*

$$\inf_{-\infty \le \omega \le \infty} |k(i\omega)| \ge d_0 > 0. \tag{8.5}$$

For instance consider the function

$$k_1(z) = z(1 - \tilde{a}e^{-\tilde{h}z}) + ae^{-hz} + b$$

with $a, b, h, \tilde{h} = const \ge 0$, and $0 < \tilde{a} < 1$. Then

$$v_1 = \frac{2(a+b)}{1 - \tilde{a}}$$

and due to Lemma 5.13 we arrive at the following result.

Corollary 5.4 *Let*

$$hv_1 < \pi/2, \tilde{h}v_1 < \pi/2 \tag{8.6}$$

and

$$d_1 := a\cos(v_1 h) + b - v_1\tilde{a}\ \sin(v_1\tilde{h}) > 0. \tag{8.7}$$

Then all the zeros of $k_1(.)$ *are in* C_- *and*

$$\inf_{-\infty\le\omega\le\infty} |k(i\omega)| \ge d_1 > 0.$$

5.9 Scalar NDEs with Positive Fundamental Solutions

Consider the scalar linear equation

$$\dot{y}(t) - a\dot{y}(t-\tilde{h}) + by(t-h) = 0 \quad (t > 0), \tag{9.1}$$

where $a, b, h, \tilde{h}$ are positive constants. The characteristic function to Eq. (9.1) is

$$k(z) = z - ze^{-\tilde{h}s}a + e^{-hs}b$$

The following result is proved in Sect. 8.4 below.

Lemma 5.13 *Let the equation*

$$s = se^{\tilde{h}s}a + e^{hs}b \tag{9.2}$$

have a positive root ψ. *Then all the zeros of* $k(z)$ *are in* C_- *and the fundamental solution (the Green function)* $Z_1(t)$ *to* (9.1) *is nonnegative. In addition,*

$$Z_1(t) \ge e^{-\psi t} \ (t \ge 0), \tag{9.3}$$

$\dot{Z}_1(t) \le 0$ *and*

$$\int_0^\infty Z_1(t)dt = \frac{1}{b}. \tag{9.4}$$

Moreover,

$$\inf_{\omega\in\mathbf{R}} |k(i\omega)| = b.$$

Note that, if (9.2) has a positive root, then necessarily $a < 1$.

Remark 5.1 If there is a positive number λ, such that

$$ae^{\tilde{h}\lambda}\lambda + be^{h\lambda} \le \lambda,$$

then due to the well-known Theorem 38.1 from the book (Krasnosel'skii and Zabreiko 1984), Eq. (9.2) has a positive root $\psi \le \lambda$.

Substitute $s = \omega c$ with a positive c into (9.2). Then

$$\omega e^{\tilde{h}c\omega} a + \frac{1}{c} e^{ch\omega} b = \omega. \tag{9.5}$$

If

$$e^{\tilde{h}c} a + \frac{1}{c} e^{ch} b \le 1, \tag{9.6}$$

then due to the just mentioned Theorem 38.1 from the book (Krasnosel'skii and Zabreiko 1984), Eq. (9.5) has a positive root which is not more than one. So (9.2) has a positive root $\psi \le c$. For example, if $c = 1/h$, then condition (9.5) takes the form

$$e^{\tilde{h}/h} a + hbe \le 1.$$

This condition is the direct generalization of the corresponding result for differential delay equations.

Minimize a function

$$f(c) = e^{\tilde{h}c} a + \frac{1}{c} e^{ch} b$$

with respect to c. It is simple to show that

$$c_0 = -\frac{bh}{2a\tilde{h}} + \sqrt{\left(\frac{bh}{2a\tilde{h}}\right)^2 + \frac{b}{a\tilde{h}}} \tag{9.7}$$

is the zero of $f'(c)$. We thus have proved

Lemma 5.14 *If*

$$e^{\tilde{h}c_0} a + \frac{1}{c_0} e^{c_0 h} b \le 1, \tag{9.8}$$

then (9.2) *has a positive root* $\psi \le c_0$.

5.10 Stability Conditions via Determinants

Consider the system

$$\dot{y}_j - \int_0^\eta \dot{y}_j(t-\tau) d\tilde{\mu}_j = \sum_{k=1}^n \int_0^\eta y_k(t-\tau) d\mu_{jk} \quad (j = 1, \dots, n), \tag{10.1}$$

where $\tilde{\mu}_j$ are nondecreasing functions and μ_{jk} $(j, k = 1, \ldots, n)$ are functions having bounded variations. In addition,

$$var(\tilde{\mu}_j) < 1 \quad (j = 1, \ldots, n). \tag{10.2}$$

The entries of the characteristic function $Q(z)$ of Eq. (10.1) are

$$k_{jk}(z) = -\int_0^\eta e^{-z\tau} d\mu_{jk} \quad (j \neq k)$$

and

$$k_{jj}(z) = z(1 - \int_0^\eta e^{-z\tau} d\tilde{\mu}_j) - \int_0^\eta e^{-z\tau} d\mu_{jj}.$$

As it was shown in Sect. 1.10, for any $n \times n$-matrix

$$M = (m_{jk})_{j,k=1}^n,$$

the inequality

$$|det\ M| \geq \prod_{j=1}^n (|m_{jj}| - \sum_{i=1, i\neq j}^n |m_{ji}|)$$

is valid, provided

$$|m_{jj}| > \sum_{i=1, i\neq j}^n |m_{ji}| \quad (j = 1, \ldots, n).$$

Hence, for a $z \in \mathbf{C}$ we have the inequality

$$|det\ Q(z)| \geq \prod_{j=1}^n (|k_{jj}(z)| - \sum_{m=1, m\neq j}^n |k_{jm}(z)|), \tag{10.3}$$

if the condition

$$|k_{jj}(z)| > \sum_{m=1, m\neq j}^n |k_{jm}(z)| \tag{10.4}$$

is fulfilled.

It can be directly checked that

$$|k_{jk}(z)| \le \sup_{\omega \in \mathbf{R}} |k_{jk}(i\omega)| \le var(\mu_{jk}) \quad (j \ne k; \ Re\ z \ge 0).$$

Let $k_{jj}(z)$ be stable: $k_{jj}(z) \ne 0$ for $Re\ z \ge 0$. Due to the minimum moduli principle for analytic functions we have

$$k_{jj}(z) \ge \inf_{\omega \in \mathbf{R}} k_{jj}(i\omega) \quad \text{for } Re\ z \ge 0.$$

Assume that

$$|k_{jj}(i\omega)| > \sum_{m=1, m \ne j}^{n} var\ (\mu_{jk}) \quad (\omega \in \mathbf{R}). \tag{10.5}$$

Then

$$|k_{jj}(z)| \ge \inf_{\omega \in \mathbf{R}} |k_{jj}(i\omega)| > \sum_{m=1, m \ne j}^{n} var(\mu_{jk}) > \sum_{m=1, m \ne j}^{n} |k_{jm}(z)| \quad (Re\ z \ge 0).$$

So $|det\ \ Q(z)| > 0$ if $Re\ z \ge 0$ and according to (10.4),

$$|det\ \ Q(i\omega)| \ge \prod_{j=1}^{n} (|k_{jj}(i\omega)| - \sum_{m=1, m \ne j}^{n} var(\mu_{jm})). \tag{10.6}$$

Due to Lemma 5.11

$$\inf_{\omega \in \mathbf{R}} |k_{jj}(i\omega)| = \inf_{-v(k_{jj}) \le \omega \le v(k_{jj})} |k_{jj}(i\omega)|, \tag{10.7}$$

where

$$v(k_{jj}) = \frac{2var(\mu_{jj})}{1 - var(\mu_j)}.$$

Taking into account (10.3) and (10.4), and applying Corollary 5.3, we arrive at our next result

Lemma 5.15 *Let all the zeros $k_{jj}(z)$ be in the open left half-plane, and the conditions* (10.2) *and*

$$\inf_{-v(k_{jj}) \le \omega \le v(k_{jj})} |k_{jj}(i\omega)| > \sum_{m=1, m \ne j}^{n} var(\mu_{jk}) \quad (j = 1, \ldots, n) \tag{10.8}$$

be fulfilled. Then system (10.1) *is exponentially stable and its characteristic determinant satisfies inequality* (10.6).

For instance, (10.1) can take the form

$$\dot{y}_j(t) - \tilde{a}_j \dot{y}_j(t - \tilde{h}_j) + \sum_{k=1}^{n} (a_{jk} y_k(t - h_j) + c_{jk} y_k(t)) = 0, \tag{10.9}$$

$(j = 1, \dots, n;\ t \geq 0)$, where $a_{jk}, c_{jk} \in \mathbf{R}$, and

$$h_j, \tilde{h}_j, a_{jj}, c_{jj} \geq 0;\ 0 < \tilde{a}_j < 1 \quad (j, k = 1, \dots, n). \tag{10.10}$$

In this case

$$k_{jk}(z) = a_{jk} e^{-zh_j} + c_{jk} \quad (j \neq k)$$

and

$$k_{jj}(z) = z(1 - \tilde{a}_j e^{-z\tilde{h}_j}) + a_{jj} e^{-zh_j} + c_{jj}.$$

Put

$$w_j = \frac{2(a_{jj} + c_{jj})}{1 - \tilde{a}_{jj}}$$

and assume that

$$w_j \max\{h_j, \tilde{h}_j\} < \frac{\pi}{2} \quad \text{and} \quad d_j := a_{jj} \cos(w_j h_j) + c_{jj} - w_j \tilde{a}_j \sin(w_j \tilde{h}_j) > 0 \tag{10.11}$$

$(j = 1, \dots, n)$. Then by Corollary 5.4, all the zeros of $k_{jj}(.)$ are in C_- and

$$\inf_{-\infty \leq \omega \leq \infty} |k_{jj}(i\omega)| \geq d_j > 0.$$

In addition, let

$$\rho_j := d_j - \sum_{m=1, m \neq j}^{n} (|a_{jm}| + |c_{jm}|) > 0 \ (j = 1, \dots, n). \tag{10.12}$$

Then according to (10.6) we get

$$|det\ Q(i\omega)| \geq \prod_{j=1}^{n} \rho_j. \tag{10.13}$$

Thus due to Lemma 5.15, we arrive at the following result.

Corollary 5.5 *Let conditions* (10.10)–(10.12) *be fulfilled. Then system* (10.9) *is exponentially stable.*

Furthermore, let us estimate the norm of the Cauchy operator to system (10.9). In the considered case we have

$$var(\tilde{R}_0) = \|\tilde{A}\|_n, var(R_0) = \|A\|_n + \|C\|_n.$$

Assuming that

$$\|\tilde{A}\|_n < 1 \tag{10.14}$$

and putting

$$\zeta_0 = \frac{2(\|A\|_n + \|C\|_n)}{1 - \|\tilde{A}\|_n},$$

we can write

$$N_2(Q(i\omega)) \le N_0(Q) \quad (|\omega| \le \zeta_0),$$

where

$$N_0(Q) := \zeta_0(\sqrt{n} + N_2(\tilde{A})) + N_2(A) + N_2(C).$$

Now Lemma 5.8 implies the estimate

$$\theta(Q) \le \hat{\theta}_d(Q); \tag{10.15}$$

here

$$\hat{\theta}_d(Q) := \frac{N_0^{n-1}(Q)}{(n-1)^{(n-1)/2} \inf_{-\zeta_0 \le \omega \le \zeta_0} |det\ (Q(i\omega))|}.$$

If condition (10.11) holds, then (10.13) is valid, and thus,

$$\theta(Q) \le \frac{N_0^{n-1}(Q)}{(n-1)^{(n-1)/2} \prod_{j=1}^n \rho_j}.$$

Now due to Theorem 5.1 we arrive at our last result in the present section.

Corollary 5.6 *Let conditions* (10.11), (10.12) *and* (10.14) *be fulfilled. Then the Cauchy operator* $\hat{Z}$ *to system* (10.9) *satisfies the inequality*

$$\|\hat{Z}\|_{L^2(0,\infty)} \le \frac{N_0^{n-1}(Q)}{(n-1)^{(n-1)/2} \prod_{j=1}^n \rho_j}.$$

5.11 NDEs with Commuting Hermitian Matrices

Let $\tilde{A}$ and A be positive definite Hermitian commuting $n \times n$-matrices having the eigenvalues $\lambda_j(A)$ and $\lambda_j(\tilde{A})$, respectively, taking with their multiplicities. Consider the equation

$$\dot{y}(t) - \tilde{A}\dot{y}(t-\tilde{h}) + Ay(t-h) = 0 \quad (t > 0). \tag{11.1}$$

In this subsection we suggest estimates for the C—and L^2—norms of the Cauchy operator to equation of (11.1), which are considerably simpler than the results of Sect. 5.7. In addition, explicit exponential stability conditions are derived.

The characteristic function to (11.1) is

$$Q(z) = z(I - \tilde{A}e^{-\tilde{h}z}) + Ae^{-hz}. \tag{11.2}$$

Rewrite (11.1) as the system in the basis of the eigenvectors of A and $\tilde{A}$, which coincide, since the matrices commute:

$$\dot{y}_j(t) - \lambda_j(\tilde{A})\dot{y}_j(t-\tilde{h}) + \lambda_j(A)y_j(t-h) = 0 \quad (j = 1, \ldots, n;\ t > 0). \tag{11.3}$$

It is assumed that each of the equations

$$s = s\lambda_k(\tilde{A})e^{s\tilde{h}} + \lambda_k(A)e^{sh}, \quad k = 1, \ldots, n, \tag{11.4}$$

have a positive root ψ_k. Then due to Lemma 5.13 the Green function $Z_j(t)$ to Eq. (11.3) is nonnegative and

$$\int_0^\infty Z_j(t)dt = \frac{1}{\lambda_j(A)}. \tag{11.5}$$

Besides, the fundamental solution $Z(t)$ to the vector Eq. (11.1) is the diagonal matrix valued function with the diagonal entries $Z_j(t)$. Put

$$x_j(t) = \int_0^t Z_j(t-s)f_j(s)ds \tag{11.6}$$

for a scalar continuous function f_j. Then

$$\sup_t \left(\sum_{k=1}^n |x_k(t)|^2\right)^{1/2} = \sup_t \left(\sum_{j=1}^n |\int_0^t Z_j(t-s)f_j(s)ds|^2\right)^{1/2}$$
$$\leq \sqrt{n}\sup_{k,s}|f_k(s)|\int_0^\infty Z_k(t)dt \leq \sqrt{n}\|f\|_{C(0,\infty)}\frac{1}{\min_k \lambda_k(A)}.$$

Moreover, by (11.5) and (11.6) we have

$$\int_0^\infty |x_j(t)|^2 dt \le \frac{1}{\lambda_j^2(A)} \int_0^\infty |f_j(t)|^2 dt.$$

Hence

$$\|x\|^2_{L^2(0,\infty)} = \int_0^\infty \sum_{k=1}^n |x_j(t)|^2 dt \le \frac{1}{\min_k \lambda_k^2(A)} \int_0^\infty \sum_{k=1}^n |f_j(t)|^2 dt.$$

We thus have proved the following

Theorem 5.3 Let $\tilde{A}$ and A be positive definite Hermitian commuting $n \times n$-matrices and Eq. (11.4) have positive roots. Then system (11.1) is exponentially stable. In addition, the Cauchy operator to (11.1) satisfies the inequalities

$$\|\hat{Z}\|_{L^2(0,\infty)} \le \frac{1}{\min_k \lambda_k(A)}$$

and

$$\|\hat{Z}\|_{C(0,\infty)} \le \frac{\sqrt{n}}{\min_k \lambda_k(A)}.$$

5.12 Autonomous NDEs with Small Principal Terms

Again consider Eq. (1.1) and put

$$H(\lambda) = I - \int_0^\eta \lambda^{-s} d\tilde{R}_0(s),$$

and

$$K(z) = Iz - \int_0^\eta exp(-zs) dR_0(s) \quad (\lambda, z \in \mathbf{C}).$$

We will say that a matrix function $F(z)$ is stable if all the zeros of $det\ F(z)$ are in the open left-half plane C_-. So $F(z)$ is invertible on the closed right-hand half plane $\overline{C}_+$

Theorem 5.4 Let $K(z)$ and $H(e^z)$ be stable and

$$var(R_0)\ var(\tilde{R}_0) \sup_{\omega\in\mathbf{R}} \|K^{-1}(i\omega)\|_n \|H^{-1}(e^{i\omega})\|_n < 1. \tag{12.1}$$

Then Eq. (1.1) is exponentially stable.

Proof Put $\hat{Q}(z) = H(e^z)K(z)$. We have

$$\hat{Q}(z) = (I - \int_0^\eta exp(-zs)d\tilde{R}_0(s))(Iz - \int_0^\eta exp(-zs)dR_0(s))$$

$$= zI - z\int_0^\eta exp(-zs)d\tilde{R}_0(s) - \int_0^\eta exp(-zs)dR_0(s)$$

$$+ \int_0^\eta exp(-zs)d\tilde{R}_0(s) \int_0^\eta exp(-zs)dR_0(s).$$

So

$$\hat{Q}(z) - Q(z) = \int_0^\eta exp(-zs)d\tilde{R}_0(s) \int_0^\eta exp(-zs)dR_0(s)).$$

But

$$\|\int_0^\eta exp(-sz)dR_0(s)\|_n \le \int_0^\eta \|dR_0(s)\|_n = var(R_0) \quad (Re\ z \ge 0) \tag{12.2}$$

and consequently

$$\|\hat{Q}(z) - Q(z)\|_n \le var(R_0)var(\tilde{R}_0)\ (Re\ z \ge 0). \tag{12.3}$$

Under the hypothesis of this theorem, $\hat{Q}(z)$ is stable. So

$$Q(z) = (I - (\hat{Q}(z) - Q(z))\hat{Q}^{-1}(z))\hat{Q}(z)\ (Re\ z \ge 0).$$

If for a given z the inequality

$$\|(\hat{Q}(z) - Q(z))Q^{-1}(z)\|_n < 1, \tag{12.4}$$

holds, then

$$Q^{-1}(z) = \hat{Q}^{-1}(z)((I - (\hat{Q}(z) - Q(z))\hat{Q}^{-1}(z))^{-1}. \tag{12.5}$$

and

$$\|Q^{-1}(i\omega)\|_n \le \frac{\|\hat{Q}^{-1}(i\omega)\|_n}{1 - \|(\hat{Q}(z) - Q(z))Q^{-1}(z)\|_n}.$$

According to (12.3) we obtain

$$\|(\hat{Q}(z) - Q(z))Q^{-1}(z)\|_n \le var(R_0)\ var(\tilde{R}_0)\|K^{-1}(z)H^{-1}(e^z)\|_n.$$

But by the maximum principle for matrix-functions (see Theorem 14.3.1 from the book (Istratescu 1981, p. 498)) we can write

$$\|K^{-1}(z)H^{-1}(e^z)\| \le \sup_{\omega \in \mathbf{R}} \|K^{-1}(i\omega)H^{-1}(e^{i\omega})\|_n.$$

Thus condition (12.1) provides inequality (12.4). So according to (12.5), $Q(z)$ is invertible on the closed right half-plane. This proves the required result. □

5.13 Stability Conditions Independent of Delays

Consider the equation

$$\dot{y} - \tilde{E}_0\dot{y} = Ay + E_1 y, \tag{13.1}$$

where A is a Hurwitzian $n \times n$-matrix:

$$\alpha(A) = \max_k Re\ \lambda_k(A) < 0, \tag{13.2}$$

$\tilde{E}_0$ is the same as in Sect. 5.1 and

$$E_1 y(t) = \int_0^\eta dR_1(\tau)y(t-\tau),$$

where R_1 is a matrix function of bounded variation.

The characteristic matrix-valued function of (13.1) is

$$Q(z) = Iz - z\int_0^\eta exp(-zs)d\tilde{R}_0(s) - A - \int_0^\eta exp(-zs)dR_1(s) \quad (z \in \mathbf{C}).$$

So

$$Q(z) = (Iz - A)(I - (Iz - A)^{-1}[z\int_0^\eta exp(-zs)d\tilde{R}_0(s) + \int_0^\eta exp(-zs)dR_1(s)]$$

for a z regular for A. Hence it follows,

$$Q^{-1}(i\omega) = [I - (Ii\omega - A)^{-1}(i\omega \int_0^\eta exp(-i\omega s)d\tilde{R}_0(s)$$

$$+ \int_0^\eta exp(-i\omega s)dR_1(s))]^{-1}(i\omega I - A)^{-1} \quad (\omega \in \mathbf{R})$$

and thus

$$\|Q^{-1}(i\omega)\|_n \leq (1 - \|(Ii\omega - A)^{-1}\|_n(|\omega|var(\tilde{R}_0) + var(R_1)))^{-1}\|(Ii\omega - A)^{-1}\|_n,$$

provided

$$\|(Ii\omega - A)^{-1}\|_n(|\omega|var(\tilde{R}_0) + var(R_1)) < 1. \tag{13.3}$$

Again assume that

$$var(\tilde{R}_0) < 1. \tag{13.4}$$

In the considered case

$$\zeta_0 = \frac{2(\|A\|_n + var(R_1))}{1 - var(\tilde{R}_0)}.$$

Thus if

$$\tau(\zeta_0) = \sup_{-\zeta_0<\omega<\zeta_0} \|(i\omega I - A)^{-1}\|_n(var(R_1) + \zeta_0 var(\tilde{R}_0)) < 1, \tag{13.5}$$

then

$$\sup_{-\zeta_0<\omega<\zeta_0} \|Q^{-1}(i\omega)\|_n \leq (1 - \tau(\zeta_0))^{-1} \sup_{-\zeta_0<\omega<\zeta_0} \|(Ii\omega - A)^{-1}\|_n < \infty.$$

Hence due to Lemma 5.6

$$\sup_{\omega\in\mathbf{R}} \|Q^{-1}(i\omega)\|_n = \sup_{-\zeta_0<\omega<\zeta_0} \|Q^{-1}(i\omega)\|_n < \infty.$$

Applying the maximum principle for matrix-functions (Theorem 14.3.1 from the book (Istratescu 1981, p. 498)), we obtain

$$\sup_{Re\ z\geq 0} \|Q^{-1}(z)\|_n < \infty.$$

Now making use Corollary 5.3, we arrive at the main result of the present section.

Theorem 5.5 Let A be a Hurwitzian matrix and conditions (13.4) and (13.5) hold. Then Eq. (13.1) is exponentially stable.

As it was shown in Sect. 2.3, one has

$$\|R_\lambda(A)\| \le \sum_{k=0}^{n-1} \frac{g^k(A)}{\sqrt{k!}\rho^{k+1}(A,\lambda)} \quad (\lambda \notin \sigma(A)). \tag{13.6}$$

Since the spectrum of A lies in C_-, we obtain

$$\rho(A, i\omega) \ge \sqrt{\alpha^2(A) + \omega^2} \ge |\alpha(A)|.$$

Hence,

$$\|(Ii\omega - A)^{-1}\| \le \sum_{k=0}^{n-1} \frac{g^k(A)}{\sqrt{k!}|\alpha(A)|^{k+1}}.$$

According to the previous theorem we thus obtain our next result.

Corollary 5.7 *Let A be a Hurwitzian matrix and the conditions* (13.4) *and*

$$\sum_{k=0}^{n-1} \frac{g^k(A)}{\sqrt{k!}|\alpha(A)|^{k+1}} (var(R_1) + \zeta_0 var(\tilde{R}_0)) < 1 \tag{13.7}$$

hold. Then Eq. (13.1) *is exponentially stable.*

5.14 Perturbations of Characteristic Determinants

Let C and C_1 be $n \times n$-matrices. Then as it is shown in Sect. 2.10,

$$|det\ C - det\ C_1| \le n\|C - C_1\| M_2^{n-1}, \tag{14.1}$$

where $M_2 := \max\{\|C\|, \|C_1\|\}$. *Throughout this section* $\|C\|$ *means the spectral norm of matrix* C.

Now let

$$Q(\lambda) = \lambda I - \lambda \int_0^\eta e^{\lambda s} d\tilde{R}_0(s) - \int_0^\eta e^{\lambda s} R_0(s)$$

and

$$Q_1(\lambda) = \lambda I - \lambda \int_0^\eta e^{\lambda s} d\tilde{R}_0(s) - \int_0^\eta e^{\lambda s} R_1(s),$$

where $R_0(s)$ and $\tilde{R}_0(s)$ are the same as above, $R_1(s) = (r^1_{ij}(s))^n_{i,j=1}$ is a real $n \times n$-matrix-valued function defined on a finite segment $[0, \eta]$ whose entries have bounded variations and finite numbers of jumps. Recall that $Q(z)$ is said to be stable if all the zeros of $det\ Q(z)$ are in C_-.

But

$$\|Q(i\omega)\| \le |\omega| + |\omega| \int_0^\eta \|d\tilde{R}_0(s)\| + \int_0^\eta \|dR_0(s)\| = |\omega|(1 + var(\tilde{R}_0)) + var(R_0)$$

and

$$\|Q_1(i\omega)\| \le |\omega|(1 + var(\tilde{R}_0)) + var(R_1) \quad (\omega \in \mathbf{R}).$$

Thus

$$\max\{\|Q(i\omega)\|, \|Q_1(i\omega)\|\} \le |\omega|(1 + var(\tilde{R})) + \max\{var(R_0), var(R_1)\}.$$

In addition,

$$\|Q(i\omega) - Q_1(i\omega)\| \le \int_0^\eta \|d(R_0(s) - R_1(s))\| = var\ (R_1 - R_0).$$

Here according to the definition of $var(R_0)$, $var\ (R_1 - R_0)$ means the spectral norm of the matrix $var\ (r^1_{ij} - r_{ij})^n_{j,k=1}$, whose entries are $var\ (r^1_{ij} - r_{ij})$.

Therefore, from (14.1) it follows

$$|det\ Q(i\omega) - det\ Q_1(i\omega)| \le \gamma(\omega), \tag{14.2}$$

where

$$\gamma(\omega) = n\ var\ (R_1 - R_0)\ [|\omega|(1 + var(\tilde{R})) + \max\{var(R_0), var(R_1)\}]^{n-1}.$$

Now Lemma 3.9 implies

Lemma 5.16 *Let $Q(\lambda)$ be stable and $|det\ Q(i\omega)| > \gamma(\omega)$ for all $\omega \in \mathbf{R}$. Then $Q_1(\lambda)$ is also stable. Moreover,*

$$|det\ Q_1(i\omega)| \ge |det\ Q(i\omega)| - \gamma(\omega) \quad (\omega \in \mathbf{R}).$$

We need also the following result.

Lemma 5.17 *Let*

$$var\ (\tilde{R}_0) < 1. \tag{14.3}$$

Then for each eigenvalue $\lambda_j(Q(i\omega))$ $(j = 1, \dots, n)$ *of* $Q(i\omega)$ *one has*

$$\inf_{-\infty \le \omega \le \infty} |\lambda_j(Q(i\omega))| = \inf_{-\zeta_0 \le \omega \le \zeta_0} |\lambda_j(Q(i\omega))|,$$

where

$$\zeta_0 = \frac{2var(R_0)}{1 - var(\tilde{R}_0)}.$$

Proof Without loss of generality, assume that $R_0(0-) = 0$. We have

$$\lambda_j(Q(i\omega)) = i\omega - \lambda_j(i\omega \int_0^\eta e^{-i\omega} \tilde{d} R_0 + \int_0^\eta e^{-i\omega} d R_0).$$

Since

$$|\lambda_j(i\omega \int_0^\eta e^{-i\omega} \tilde{d} R_0 + \int_0^\eta e^{-i\omega} d R_0)|, \le |\omega| var(\tilde{R}_0) + var(R_0),$$

this yields the inequality

$$|\lambda_j(Q(i\omega))| \ge |\omega|(1 - var(\tilde{R}_0)) - var(R_0) \ge var(R_0) \quad (|\omega| \ge \zeta_0).$$

But

$$|\lambda_j(Q(0))| = |\lambda_j(\int_0^\eta d R_0)| = |\lambda_j(R_0(\eta))| \le \|R_0(\eta)\|.$$

Taking into account that $\|R_0(\eta)\| \le var(R_0)$, we can assert that the minimum of $|\lambda_j(Q(i\omega))|$ is attained on $[-\zeta_0, \zeta_0]$. □

Corollary 5.8 *If condition* (14.3) *is fulfilled, then the equality*

$$\inf_{-\infty \le \omega \le \infty} |det \ Q(i\omega)| = \inf_{-\zeta_0 \le \omega \le \zeta_0} |det \ Q(i\omega)|$$

is valid.

Furthermore, put

$$\zeta_2 := \frac{2 \max\{var\ (R_0), var\ (R_1)\}}{1 - var(\tilde{R}_0)}.$$

Then

$$\gamma(\omega) \le \hat{\gamma} \quad (|\omega| \le \zeta_2),$$

where

$$\hat{\gamma} := var\,(R_1 - R_0)\ \ (\zeta_2(1 + var\,(\tilde{R}_0)) + \max\{var\,(R_0), var\,(R_1)\})^{n-1}.$$

Now Lemma 5.16 and Corollary 5.8 imply the main result of the present section.

Theorem 5.6 Let condition (14.3) hold and $Q(\lambda)$ be stable. In addition, let

$$\inf_{-\zeta_2 \le \omega \le \zeta_2} |det\ Q(i\omega)| > \hat{\gamma}.$$

Then $Q_1(z)$ is also stable and

$$|det\ Q_1(i\omega)| \ge \inf_{-\zeta_2 \le \omega \le \zeta_2} |det\ Q(i\omega)| - \hat{\gamma} \quad (\omega \in \mathbf{R}).$$

5.15 Comments

The material of this chapter is based on the paper (Gil' 2012c).

Following Hale (1977, Sect. 1.7), let us discuss the nature of solutions of the neutral differential difference equations. The model scalar nonhomogeneous equation is

$$\dot{x}(t) - C\dot{x}(t - \eta) = Ax(t) + Bx(t - \eta) + f(t)\ (t \ge 0), \tag{15.1}$$

where A, B, C, and η are constants with $\eta > 0$, $C \ne 0$ and f is a continuous function. The corresponding homogeneous equation is

$$\dot{x}(f) - C\dot{x}(t - \eta) = Ax(t) + Bx(t - \eta). \tag{15.2}$$

In the retarded case, the initial space did not play an important role in the initial-value problem since the solution becomes continuously differentiable for $t > 0$ and, in particular, it is continuous. For Eq. (15.1), it is a little more difficult to define the concept of a solution and the appropriate space of initial data. We can see that such smoothing of the solution does not occur with increasing t.

Following the arguments of Lemma 5.2 one can show that there always exists a unique solution $x(t)$ of Eq. (15.1) through a given continuously differentiable function ϕ defined on $[-\eta, 0]$. Besides $x(t)$ is a continuous function defined on $[-\eta, \infty)$ with $x(t) = \phi(t)$ for $t \in [-\eta, 0]$, continuously differentiable, except at the points $k\eta$ $(k = 0, 1, 2, \ldots)$ and satisfying Eq. (15.1) except at these points. Equation (15.1) implies that the solution can never have more derivatives than the initial function. Also, the solution will in general have a discontinuous derivative at $k\eta$, $k = 0, 1, 2.$ In fact, suppose

$$\dot{\phi}(0) \ne C\dot{\phi}(-\eta) + A\phi(0) + B\phi(-\eta) + f(0).$$

Then the solution $\dot{x}(t)$ has a discontinuous derivative at $t = 0$. From Eq. (15.1), this implies that $\dot{x}(t)$ is discontinuous for $t = \eta$. The same reasoning applies for all values $k\eta$. On the other hand, if the "sewing condition"

$$\dot{\phi}(0) = C\dot{\phi}(-\eta) + A\phi(0) + B\phi(-\eta) + f(0) \tag{15.3}$$

holds, then the solution has a continuous derivative at $t = 0$. Equation (15.1) can again be used to show that $x(t)$ has a continuous derivative for all $t > \eta$. Consequently, relation (15.3) is necessary and sufficient for the solution of (15.1) to have a continuous derivative for all $t > \eta$.

Furthermore, the stability of various concrete classes of the autonomous neutral type functional-differential equations has been intensively investigated, cf. the well-known books (Hale 1977; Hale and Lunel 1993; Kolmanovskii and Myshkis 1999 and Kolmanovskii and Nosov 1986).

In the case of discrete delays, the stability analysis can be reduced to an analysis of a quasipolynomial of the form

$$D(z) = \sum_{i=0}^{m}\sum_{j=0}^{n} a_{ij} z^i exp(b_j z)$$

with positive b_j. In particular, a quasipolynomial with commensurable delays can be represented as

$$D(z) = \sum_{i=0}^{m}\sum_{j=0}^{n} a_{ij} z^i exp(jz).$$

The most general results on the roots of that quasipolynomial are due to (Pontryagin 1955) (see also (Hale 1977, Appendix)). We state the results of Pontryagin (Theorems 5.7–5.10) without proofs. Put

$$P(z, w) = \sum_{i=0}^{m}\sum_{j=0}^{n} a_{ij} z^i w^j$$

Then $D(z) = P(z, e^z)$. A term $a_{i_0 j_0} z_0^i w_0^j$ is called the principal term of the polynomial $P(z, w)$, if $a_{i_0 j_0} \neq 0$, and if, for each other term $a_{ij} z^i w^j$ with $a_{ij} \neq 0$ we have either $j < j_0, i < i_0$, or $j = j_0, i < i_0$, or $j < j_0, i = i_0$.

Theorem 5.7 If the polynomial $P(z, w)$ has no a principal term, then the equation $P(z, e^z) = 0$ has an infinite number of zeros with arbitrarily large positive real parts.

The basic results for applications are the next three theorems.

Theorem 5.8 Let $D(z) = P(z, e^z)$ and suppose $P(z, w)$ is a polynomial with a principal term $a_{rs} z^r w^s$. All of the zeros of $D(z)$ have negative real parts if and only if

(i) The complex vector $D(i\omega)$ rotates in the positive direction with a positive velocity for ω ranging in $(-\infty, \infty)$.
(ii) For $\omega \in [-2k\pi, 2k\pi]$, $k \geq 0$ is an integer, there is an $\varepsilon_k \to 0$ as $k \to \infty$ such that $D(i\omega)$ subtends an angle $4k\pi s + \pi r + \varepsilon_k$.

Theorem 5.9 Let $D(z) = P(z, e^z)$, where $P(z, w)$ is a polynomial with a principal term. Suppose $D(i\omega)$ $(\omega \in \mathbf{R})$ is separated into its real and imaginary parts, $D(i\omega) = F(\omega) + iG(\omega)$. If all the zeros of $D(z)$ have negative real parts, then all the zeros of $F(\omega)$ and $G(\omega)$ are real, simple, alternate and

$$G'(\omega)F(\omega) - F'(\omega)G(\omega) > 0 \tag{15.4}$$

for every $\omega \in \mathbf{R}$. Conversely, all the zeros of $P(z, e^z)$ have negative real parts provided that either of the following conditions are satisfied:

(i) All the zeros of $F(\omega)$ and $G(\omega)$ are real, simple, alternate and inequality (15.4) holds for at last one ω.
(ii) All the zeros of $F(\omega)$ are real and, for each zero inequality (15.4) is satisfied.
(iii) All the zeros of $G(\omega)$ are real and, for each zero inequality (15.4) is satisfied.

Now suppose that $f(z, u, v)$ is a polynomial in z, u, v with real coefficients, which has the form

$$f(z, u, v) = \sum_{m=0}^{r} \sum_{n=0}^{s} z^m \phi_m^{(n)}(u, v) \tag{15.5}$$

where $\phi_m^{(n)}(u, v)$ is a homogeneous polynomial of degree n in u, v. *The principal term* in the polynomial $f(z, u, v)$ is the term $z^r \phi_r^{(s)}(u, v)$, for which either $r > m,\ s > n$, or $r = m,\ s > n$, or $r > m,\ s = n$ for all other terms in (15.5).

Let $z^r \phi_r^{(s)}(u, v)$ denote the principal term of $f(z, u, v)$ in (15.5), let $\phi_*^{(s)}(u, v)$ denote the coefficient of z^r in $f(z, u, v)$,

$$\phi_*^{(s)}(u, v) = \sum_{n=0}^{s} \phi_r^{(n)}(u, v),$$

and let

$$\Phi_*^{(s)}(z) = \phi_*^{(s)}(\cos z, \sin z).$$

Theorem 5.10 Let $f(z, u, v)$ be a polynomial with a principal term

$$z^r \phi_r^{(s)}(u, v).$$

If ε is such that $\Phi_*^{(s)}(\varepsilon + i\omega) \neq 0,\ \omega \in \mathbf{R}$, then for sufficiently large integers k, the function $F(z) = f(z, cosz, sinz)$ will have exactly $4ks + r$ zeros in the strip

$-2k\pi + \varepsilon \leq Re\, z \leq 2k\pi + \varepsilon$. Consequently, the function $F(z)$ will have only real roots if and only if, for sufficiently large integers k, it has exactly $4ks + r$ zeros in the strip $-2k\pi + \varepsilon \leq Re\, z \leq 2k\pi + \varepsilon$.

About the recent results on the stability of autonomous NDEs see for instance the very interesting papers (Cahlon and Schidt 2011; Demidenko 2009; Han 2008; Park and Won 2000; Walther 2011).

Chapter 6
Linear Time-Variant NDEs

This chapter is devoted to the stability analysis of linear vector time-variant (non-autonomous) neutral type functional differential equations. In particular, we extend the Bohl-Perron principle to a class of neutral type functional differential equations. Namely, it is proved that the homogeneous equation is exponentially stable, provided the corresponding non-homogeneous equation with the zero initial condition and an arbitrary bounded free term has a bounded solution. The required result at once

We also establish the integral version of the generalized Bohl-Perron principle for NDEs, i.e. we show that the homogeneous equation is exponentially stable, if the corresponding non-homogeneous equation with the zero initial condition and an arbitrary free term from $L^p(0, \infty) = L^p([0, \infty), \mathbf{C}^n)$, has a solution belonging to $L^p(0, \infty)$.

As applications of these principles, the stability conditions are derived for time-variant systems with discrete delays, which are close to autonomous systems.

In addition, we investigate time-variant systems with small principal operators and obtain stability conditions independent of delay in the non-autonomous case.

6.1 Existence Results for Linear NDEs

Consider the problem

$$\frac{d}{dt}\left[x(t) - \int_0^\eta d_\tau \tilde{R}(t,\tau)x(t-\tau)\right] = \int_0^\eta d_\tau R(t,\tau)x(t-\tau) + f(t) \quad (t > 0;\ 0 < \eta < \infty), \tag{1.1}$$

$$x(t) = 0 \quad (-\eta \le t \le 0), \tag{1.2}$$

M.I. Gil', *Stability of Neutral Functional Differential Equations*,
Atlantis Studies in Differential Equations 3, DOI 10.2991/978-94-6239-091-1_6

where f is a function defined on $[0, \infty)$ with values in $\mathbf{C}^n$ and integrable on each finite interval $(0, T)$; $R(t, \tau) = (r_{ij}(t, \tau))_{i,j=1}^n$ and $\tilde{R}(t, \tau) = (\tilde{r}_{ij}(t, \tau))_{i,j=1}^n$ are real $n \times n$-matrix-valued functions defined on $[0, \infty) \times [0, \eta]$, whose entries have uniformly bounded variations in τ:

$$v_{jk} := \sup_{t \geq 0} var\ r_{ij}(t, .) < \infty \tag{1.3a}$$

and

$$\tilde{v}_{jk} := \sup_{t \geq 0} var\ \tilde{r}_{ij}(t, .) < \infty \quad (j, k = 1, \ldots, n). \tag{1.3b}$$

Besides, $R(t, \tau)$ is continuous in t and $\tilde{R}(t, \tau)$ is continuously differentiable in t. In addition, $R(t, \tau)$ and $\tilde{R}(t, \tau)$ have finite numbers of jumps in τ and $\tilde{R}(t, \tau)$ does not have a jump at $\tau = 0$ for any $t \geq 0$.

The integrals in (1.1) are understood as the Lebesgue-Stieltjes integrals.

A solution of problem (1.1), (1.2) is an absolutely continuous vector valued function $x(t)$ defined on $[-\eta, \infty)$ and satisfying (1.2) and the equation

$$x(t) - \int_0^\eta d_\tau \tilde{R}(t, \tau)x(t - \tau) = \int_0^t f(t_1)dt_1 + \int_0^t \int_0^\eta d_\tau R(t, \tau)x(t_1 - \tau)dt_1 \ (t \geq 0). \tag{1.4}$$

Put

$$Su(t) = \int_0^{\min\{\eta, t\}} d_\tau \tilde{R}(t, \tau)u(t - \tau) \ \ (t \geq 0) \text{ and } Su(t) = 0 \ \ (t < 0)$$

$(u \in L^\infty(0, T))$. In addition,

$$Eu(t) = \int_0^\eta d_\tau R(t, \tau)u(t - \tau).$$

So (1.1) and (1.4) can be written as

$$\frac{d}{dt}[x - Sx] = Ex + f$$

and

$$x(t) - Sx(t) = \int_0^t Ex(s)ds + \int_0^t f(s)ds,$$

respectively.

Lemma 6.1 *Assume that conditions* (1.3a), (1.3b) *hold, and the spectral radius of operator S in $L^\infty(0,T)$ ($T < \infty$) is less than one, and a solution of* (1.1), (1.2) *admits a priory estimate for $\|x\|_{C(0,T)}$. Then* (1.1) *possesses a solution.*

Proof Introduce the operator

$$Ju(t) = \int_0^t u(s)ds.$$

Then (1.4) takes the form

$$x = (I-S)^{-1}J(Ex+f).$$

The operator J is compact in $L^\infty(0,T)$, the operator $(I-S)^{-1}$ is bounded in $L^\infty(0,T)$; so $(I-S)^{-1}J$ is compact in $L^\infty(0,T)$. But $(I-S)^{-1}J$ maps $C(0,T)$ into itself. So $(I-S)^{-1}J$ is compact in $C(0,T)$. In addition, E is bounded in $C(0,T)$. Consequently, the operator $(I-S)^{-1}JE$ is compact in $C(0,T)$. Now the required result is due to the Leray-Schauder principle (Krasnosel'skij and Zabreiko 1984). □

Set $V(R) := \|(v_{jk})\|_n$-the spectral (operator) norm of matrix $(v_{jk})_{j,k=1}^n$. Similarly, $V(\tilde{R}) := \|(\tilde{v}_{jk})\|_n$.

Lemma 6.2 *Let the conditions* (1.3a), (1.3b) *and*

$$V(\tilde{R}) < 1 \tag{1.5}$$

hold. Then a solution of problem (1.1), (1.2) *satisfies the a priory estimate*

$$\|x(t)\|_n \le \frac{c_1\ \|f\|_{L^1(0,t)}}{1-V(\tilde{R})} exp\left[\frac{tV(R)}{1-V(\tilde{R})}\right] \quad (t \ge 0), \tag{1.6}$$

where c_1 is a constant independent of f.

Proof For a vector function $v(t) = (v_k(t))$ again introduce the norm

$$\|v\|_{\hat{C}([a,b],\mathbf{C}^n)} = [\sum_{k=1}^n \sup_{a\le t\le b} |v_k(t)|^2]^{1/2},$$

(see Sect. 1.8). Besides,

$$\|v\|_{C([a,b],\mathbf{C}^n)} = \sup_{a\le t\le b} [\sum_{k=1}^n |v_k(t)|^2]^{1/2} \le \|v\|_{\hat{C}([a,b],\mathbf{C}^n)}$$

for all $-\infty < a < b \leq \infty$. Due to Lemma 1.11 we have

$$\|Sv\|_{\hat{C}([0,T],\mathbf{C}^n)} \leq V(\tilde{R})\|v\|_{\hat{C}([-\eta,T],\mathbf{C}^n)}$$

and

$$\|Ev\|_{\hat{C}([0,T],\mathbf{C}^n)} \leq V(R)\|v\|_{\hat{C}([-\eta,T],\mathbf{C}^n)}.$$

Now (1.4) implies

$$\|x\|_{\hat{C}([0,T],\mathbf{C}^n)} \leq const\|f\|_{L^1(0,T)} + V(\tilde{R})\|x\|_{\hat{C}([0,T],\mathbf{C}^n)} + V(R)\int_0^t \|x\|_{\hat{C}([0,s],\mathbf{C}^n)}ds.$$

Put $m(t) = \|x\|_{\hat{C}([0,t],\mathbf{C}^n)}$. Then we can write

$$m(t) \leq const\|f\|_{L^1(0,t)} + V(\tilde{R})m(t) + V(R)\int_0^t m(s)ds.$$

Hence,

$$m(t) \leq \frac{1}{1-V(\tilde{R})}[const\ \|f\|_{L^1(0,t)} + V(R)\int_0^t m(s)ds].$$

Now the Gronwall lemma yields the inequality

$$m(t) \leq \frac{const\ \|f\|_{L^1(0,t)}}{1-V(\tilde{R})} exp\left[\frac{V(R)t}{1-V(\tilde{R})}\right] \quad (t \geq 0).$$

This yields the required result. □

The latter lemma and the Leray-Schauder principle (Krasnosel'skii and Zabreiko, 1984) yield.

Corollary 6.1 *Let conditions* (1.3a) *and* (1.5) *hold. Then problem* (1.1), (1.2) *possesses a unique solution.*

Now consider the homogeneous problem

$$\frac{d}{dt}\left[y(t) - \int_0^\eta d_\tau\tilde{R}(t,\tau)y(t-\tau)\right] - \int_0^\eta d_\tau R(t,\tau)y(t-\tau) = 0 \quad (t > 0), \tag{1.7}$$

$$y(t) = \phi(t) \quad (-\eta \leq t \leq 0), \tag{1.8}$$

where $\phi \in C^1(-\eta, 0)$ is given. A solution of (1.7), (1.8) is an absolutely continuous vector valued function $y(t)$ defined on $[-\eta, \infty)$ and satisfying (1.8) and the equation

$$y(t) - \int_0^\eta d_\tau \tilde{R}(t, \tau) y(t-\tau) = \phi(0) - \int_0^\eta d_\tau \tilde{R}(0, \tau)\phi(-\tau)$$

$$+ \int_0^t \int_0^\eta d_\tau R(t_1, \tau) y(t_1 - \tau) dt_1 \ (t \geq 0). \tag{1.9}$$

Lemma 6.3 *Let the conditions* (1.3a), (1.3b) *and* (1.5) *hold. Then problem* (1.7), (1.8) *has a unique solution.*

Proof Put

$$\hat{\phi}(t) = \begin{cases} \phi(0) & \text{if } t > 0, \\ \phi(t) & \text{if } -\eta \leq \phi(t) \leq 0 \end{cases}$$

and substitute $x_0(t) = y(t) - \hat{\phi}(t)$ into (1.1). Then we obtain the equation

$$\frac{d}{dt}[x_0(t) - \int_0^\eta d_\tau \tilde{R}(t, \tau) x_0(t-\tau)] = \int_0^\eta d_\tau R(t, \tau) x_0(t-\tau) + f_0(t),$$

where

$$f_0(t) = -\frac{d\hat{\phi}(t)}{dt} + \frac{d}{dt}\int_0^\eta d_\tau \tilde{R}(t, \tau)\hat{\phi}(t-\tau) + \int_0^\eta d_\tau R(t, \tau)\hat{\phi}(t-\tau).$$

So problem (1.7), (1.8) is reduced to problem (1.1), (1.2). Besides, f_0 is locally integrable, since $R(t, \tau)$ is continuously differentiable in $t \geq 0$, and $d\hat{\phi}(t-\tau)/dt = d\phi(t-\tau)/dt$ for $t \leq \tau$, and $d\hat{\phi}(t-\tau)/dt = 0$ for $t > \tau$. Now the required result is due to Corollary 6.1. □

About the existence results for linear NDEs under conditions of the type (1.3a), (1.3b) see also the well-known Theorem 9.1.1 from the book by Hale and Lunel (1993, p. 256).

As in the autonomous case, Eq. (2.1) is said to be exponentially stable if there are positive constants M and α independent on ϕ such that

$$\|y(t)\|_n \leq M e^{-\alpha t} \|\phi\|_{C^1(-\eta, 0)} \ \ (t \geq 0)$$

for any solution $y(t)$ of problem (1.7), (1.8).

Let $\hat{Z}$ be an operator defined by $\hat{Z}f = x$, where x is a solution of problem (1.1), (1.2). Then $\hat{Z}$ will be called *the Cauchy operator to Eq.* (1.1) *and to Eq.* (1.7). That is, $\hat{Z}$ solves problem (1.1), (1.2).

6.2 The Generalized Bohl-Perron Principle

In the present section we extend the Bohl-Perron principle to a class of neutral type functional differential equations.

Let $\tilde{A}_j(t)$ $(t \geq 0;\ j = 1, \ldots, m_1)$ be continuously differentiable $n \times n$-matrices; $A_k(t)$ $(t \geq 0;\ k = 1, \ldots, m)$ be $n \times n$-matrices defined and continuous on the positive half-line. In addition, $\tilde{A}(t, \tau)$ $(t \geq 0;\ \tau \in [0, \eta])$ is an $n \times n$-matrix continuously differentiable in t for each τ; $A(t, \tau)$ $(t \geq 0;\ \tau \in [0, \eta])$ is an $n \times n$-matrix continuous in t for each τ; $\tilde{A}(t, \tau)$, $\tilde{A}'_t(t, \tau)$ and $A(t, \tau)$ are integrable in τ on $[0, \eta]$.

At $u \in C(-\eta, \infty)$ define the operators E_2, E_1 by

$$(E_1u)(t) = \sum_{k=1}^{m_1} \tilde{A}_k(t)u(t - \tilde{h}_k) + \int_0^{\eta} \tilde{A}(t, s)u(t - s)ds \quad (t \geq 0),$$

and

$$(E_2u)(t) = \sum_{k=1}^{m} A_k(t)u(t - h_k(t)) + \int_0^{\eta} A(t, s)u(t - s)ds,$$

where

$$0 < \tilde{h}_1 < \cdots < \tilde{h}_{m_1} \leq \eta \quad (m_1 < \infty)$$

are constants, $h_j(t)$ are real continuous functions, such that

$$0 \leq h_j(t) \leq \eta \quad (t \geq 0;\ j = 1, \ldots, m).$$

Our main object in this section is the equation

$$\frac{d}{dt}[y(t) - (E_1y)(t)] = (E_2y)(t) \quad (t > 0) \tag{2.1}$$

with the initial condition

$$y(t) = \phi(t) \quad (-\eta \leq t \leq 0) \tag{2.2}$$

for a given $\phi \in C^1(-\eta, 0)$.

We consider also the non-homogeneous equation

$$\frac{d}{dt}[x(t) - (E_1 x)(t)] = (E_2 x)(t) + f(t) \quad (t > 0) \tag{2.3}$$

with a given vector function $f \in C(0, \infty)$ and the zero initial condition

$$x(t) = 0 \quad (-\eta \le t \le 0). \tag{2.4}$$

Put

$$V_0 := \sup_{t \ge 0} \left[\sum_{k=1}^{m} \|A_k(t)\|_n + \int_0^{\eta} \|A(t, s)\|_n ds \right]$$

and

$$\hat{V} := \sup_{t \ge 0} \left[\sum_{k=1}^{m_1} \|\tilde{A}_k'(t)\|_n + \int_0^{\eta} \|\tilde{A}_t'(t, s)\|_n ds \right].$$

It is assumed that the conditions

$$\hat{V} < \infty, V_0 < \infty \text{ and } V_1 := \left[\sum_{k=1}^{m_1} \sup_{t \ge 0} \|\tilde{A}_k(t)\|_n + \int_0^{\eta} \sup_{t \ge 0} \|\tilde{A}(t, s)\|_n ds \right] < 1 \tag{2.5}$$

hold. According to the notations of the previous section it is not hard to check that $V(R) \le V_0$ and $V(\tilde{R}) \le V_1$, see also Lemma 6.4.

Solutions of problems (2.1), (2.2), (2.3) and (2.4) are defined as in Sect. 6.1. The existence and uniqueness of solutions under condition (2.5) are due to Corollary 6.1 and Lemma 6.3.

Now we are in a position to formulate our next result.

Theorem 6.1 *Let condition* (2.5) *hold. If, in addition, a solution $x(t)$ of problem* (2.3), (2.4) *is bounded on* $[0, \infty)$ (*that is,* $x \in C(0, \infty)$) *for any* $f \in C(0, \infty)$, *then Eq.* (2.1) *is exponentially stable.*

This theorem is proved in the next section.

Remark 6.1 We consider Eq. (2.1) for the sake of the simplicity. In the arguments of the proof of Theorem 6.1 one can replace the operator E_2 by the operator E defined in the previous section, provided condition (1.3a) holds.

6.3 Proof of Theorem 6.1

Rewrite (2.1) as

$$\dot{y}(t) - (E_1\dot{y})(t) = (E_1'y)(t) + (E_2y)(t) \quad (t \geq 0), \tag{3.1}$$

where

$$(E_1'y)(t) = \sum_{k=1}^{m_1} \tilde{A}_k'(t)y(t-\tilde{h}_k) + \int_0^\eta \tilde{A}_t'(t,s)y(t-s)ds.$$

Lemma 6.4 *For any $T > 0$ the inequalities*

$$\|E_1u\|_{C(0,T)} \leq V_1\|u\|_{C(-\eta,T)}, \ \|E_2u\|_{C(0,T)} \leq V_0\|u\|_{C(-\eta,T)}$$

and

$$\|E_1'u\|_{C(0,T)} \leq \hat{V}\|u\|_{C(-\eta,T)} \quad (u \in C(-\eta,T))$$

are valid.

Proof Let $u \in C(-\eta, T)$. We have

$$\begin{aligned}\|(E_1u)(t)\|_n &\leq \sum_{k=1}^{m_1} \|\tilde{A}_k(t)u(t-\tilde{h}_k)\|_n + \int_0^\eta \|\tilde{A}(t,s)u(t-s)\|_n ds \\ &\leq \|u\|_{C(-\eta,T)}(\sum_{k=1}^{m_1} \|\tilde{A}_k(t)\|_n \\ &\quad + \int_0^\eta \|\tilde{A}(t,s)\|_n ds) \leq V_1\|u\|_{C(-\eta,T)} \quad (0 \leq t \leq T).\end{aligned}$$

Similarly the other inequalities can be proved. This proves the lemma. □

Lemma 6.5 *If for any $f \in C(0,\infty)$ a solution of problem* (2.3), (2.4) *is in $C(0,\infty)$, and conditions* (2.5) *hold, then any solution of problem* (2.1), (2.2) *is in $C(-\eta,\infty)$.*

Proof Let $y(t)$ be a solution of problem (2.1), (2.2). Again put

$$\hat{\phi}(t) = \begin{cases} \phi(0) & \text{if } t \geq 0, \\ \phi(t) & \text{if } -\eta \leq t \leq 0 \end{cases}$$

and $x_0(t) = y(t) - \hat{\phi}(t)$. We can write $d\hat{\phi}(t)/dt = 0 \ \ (t \geq 0)$ and

$$\frac{d}{dt}[x_0(t) - (E_1 x_0)(t)] = (E_2 x_0)(t) + f_0(t) \quad (t > 0),$$

where

$$f_0(t) = \frac{d(E_1\hat{\phi})(t)}{dt} + (E_2\hat{\phi})(t) = (E_1\frac{d\hat{\phi}}{dt})(t) + (E_1'\hat{\phi})(t) + (E_2\hat{\phi})(t).$$

Besides, (2.4) holds with $x(t) = x_0(t)$. Since $\hat{\phi} \in C^1(-\eta, \infty)$, by the previous lemma we have $f_0 \in C(-\eta, \infty)$. Due to the hypothesis of this lemma, $x_0 \in C(0, \infty)$. Thus, $y = x_0 + \hat{\phi} \in C(-\eta, \infty)$. As claimed. □

Lemma 6.6 *Let conditions* (2.5) *hold. Then for any solution x of problem* (2.3), (2.4) *and all* $T > 0$*, one has*

$$\|\dot{x}\|_{C(0,T)} \leq \frac{V_0\|x\|_{C(0,T)} + \|f\|_{C(0,T)}}{1 - V_1}.$$

Proof From (3.1), by Lemma 6.4 we have,

$$\|\dot{x}\|_{C(0,T)} \leq V_1\|\dot{x}\|_{C(0,T)} + V_0\|x\|_{C(0,T)} + \|f\|_{C(0,T)}.$$

Hence the condition $V_1 < 1$ implies the required result. □

Proof of Theorem 6.1 Substituting

$$y(t) = y_\varepsilon(t)e^{-\varepsilon t} \tag{3.2}$$

with an $\varepsilon > 0$ into (3.1), we obtain the equation

$$\dot{y}_\varepsilon - \varepsilon y_\varepsilon - E_{\varepsilon,1}\dot{y}_\varepsilon + \varepsilon E_{\varepsilon,1}y_\varepsilon = (E_{\varepsilon,1}' + E_{\varepsilon,0})y_\varepsilon, \tag{3.3}$$

where

$$(E_{\varepsilon,1}f)(t) = \sum_{k=1}^{m_1} e^{\tilde{h}_k\varepsilon}\tilde{A}_k(t)f(t-\tilde{h}_k) + \int_0^\eta e^{s\varepsilon}A(t,s)f(t-s)ds,$$

$$(E_{\varepsilon,0}f)(t) = \sum_{k=1}^{m} A_k(t)e^{h_k(t)\varepsilon}f(t-h_k(t)) + \int_0^\eta A(t,s)e^{s\varepsilon}f(t-s)ds$$

and

$$(E'_{\varepsilon,1}f)(t) = \sum_{k=1}^{m_1} e^{\tilde{h}_k\varepsilon}\tilde{A}'_k(t)f(t-\tilde{h}_k) + \int_0^\eta e^{s\varepsilon}A'_t(t,s)f(t-s)ds.$$

Rewrite (3.3) as

$$\frac{d}{dt}[y_\varepsilon - E_{\varepsilon,1}y_\varepsilon] = E_{\varepsilon,0}y_\varepsilon + W_\varepsilon y_\varepsilon, \tag{3.4}$$

where

$$W_\varepsilon := \varepsilon I - \varepsilon E_{\varepsilon,1}.$$

Furthermore, introduce in $C(0,\infty)$ the Cauchy operator $\hat{Z} : f \to x$ where $x(t)$ is the solution of problem (2.3), (2.4). That is, $\hat{Z}$ solves problem (2.3), (2.4).

By the hypothesis of the theorem, we have

$$x = \hat{Z}f \in C(0,\infty) \text{ for any } f \in C(0,\infty).$$

So $\hat{Z}$ is defined on the whole space $C(0,\infty)$. It is closed, since problem (2.3), (2.4) under conditions (2.5) has a unique solution due to Corollary 6.1. Therefore $\hat{Z} : C(0,\infty) \to C(0,\infty)$ is bounded according to the Closed Graph Theorem (Theorem 1.3 of the present book). So the norm $\|\hat{Z}\|_{C(0,\infty)}$ is finite. Consider now the equation

$$\frac{d}{dt}[x_\varepsilon - E_{\varepsilon,1}x_\varepsilon] = E_{\varepsilon,0}x_\varepsilon + W_\varepsilon x_\varepsilon + f \tag{3.5}$$

with the zero initial conditions. Subtract (2.3) from (3.5), with $w(t) = x_\varepsilon(t) - x(t)$, where x and x_ε are solutions of problems (2.3), (2.4), (3.5) and (2.4), respectively. Then

$$\frac{d}{dt}[w - E_1 w] = E_2 w + F_\varepsilon \tag{3.6}$$

where

$$F_\varepsilon = W_\varepsilon + (E_{\varepsilon,0} - E_2)x_\varepsilon \frac{d}{dt}(E_{\varepsilon,1} - E_1)x_\varepsilon. \tag{3.7}$$

It is simple to check that $W_\varepsilon \to 0$, $E'_{\varepsilon,1} \to E'_1$ and $E_{\varepsilon,1} \to E_1$ in the operator norm of $C(0,\infty)$ as $\varepsilon \to 0$.

For the brevity in this proof put $\|.\|_{C(0,T)} = |.|_T$ for a finite $T > 0$. We have

$$\frac{d}{dt}(E_{\varepsilon,1} - E_1)x_\varepsilon = (E_{\varepsilon,1} - E_1)\dot{x}_\varepsilon + (E'_{\varepsilon,1} - E'_1)x_\varepsilon.$$

So

$$|\frac{d}{dt}(E_{\varepsilon,1} - E_1)x_\varepsilon|_T \leq a_1(\varepsilon)|\dot{x}_\varepsilon|_T + a_2(\varepsilon)|x_\varepsilon|_T, \tag{3.8}$$

where

$$a_j(\varepsilon) \to 0 \text{ as } \varepsilon \to 0$$

for an index j. But according to Lemma 6.4, for a sufficiently small $\varepsilon > 0$, we have

$$|E_{\varepsilon,1}x_\varepsilon|_T \leq e^{\varepsilon\eta}V_1|x_\varepsilon|_T \text{ with } e^{\varepsilon\eta}V_1 < 1. \tag{3.9}$$

Due Lemma 6.6, from (3.5), the inequality

$$|\dot{x}_\varepsilon|_T \leq (1 - e^{\varepsilon\eta}V_1)^{-1}(|E'_{\varepsilon,1} + W_\varepsilon|_T|x_\varepsilon|_T + |f|_T)$$

is fulfilled. Take into account that $W_\varepsilon \to 0$ and $E'_{\varepsilon,1} \to E'_1$. So $|E'_{\varepsilon,1}|_T$ is bounded uniformly with respect to ε and $T > 0$, for a sufficiently small $\varepsilon_0 > 0$, and therefore there is a constant c_1, such that

$$|\dot{x}_\varepsilon|_T \leq c_1(|x_\varepsilon|_T + |f|_T) \quad (\varepsilon < \varepsilon_0;\ T > 0).$$

Now (3.8) implies

$$|\frac{d}{dt}(E_{\varepsilon,1} - E_1)x_\varepsilon|_T \leq a_1(\varepsilon)c_1(|x_\varepsilon|_T + |f|_T) + a_2(\varepsilon)|x_\varepsilon|_T \leq a_3(\varepsilon)(|x_\varepsilon|_T + |f|_T). \tag{3.10}$$

Furthermore

$$|(E_{\varepsilon,0} - E_{\varepsilon,0})x_\varepsilon|_T \leq a_4(\varepsilon)|x_\varepsilon|_T.$$

Thus from (3.10) we deduce that

$$|F_\varepsilon|_T \leq a(\varepsilon)(|x_\varepsilon|_T + |f|_T),$$

uniformly in $T > 0$. By (3.6) $x_\varepsilon - x = \hat{Z}F_\varepsilon$. So

$$|x_\varepsilon - x|_T \leq \|\hat{Z}\|_{C(0,\infty)}a(\varepsilon)(|x_\varepsilon|_T + |f|_T).$$

For a sufficiently small ε, we have

$$q(\varepsilon) := \|\hat{Z}\|_{C(0,\infty)}a(\varepsilon) < 1.$$

Thus

$$|x_\varepsilon|_T \leq (1 - q(\varepsilon))^{-1}(|x|_T + \|\hat{Z}\|_{C(0,\infty)}a(\varepsilon)|f|_T).$$

By the hypothesis of the present theorem, $x(t)$ is bounded on $[0, \infty)$. This gives us the inequality

$$|x_\varepsilon|_T \le (1 - q(\varepsilon))^{-1}(\|x\|_{C(0,\infty)} + a_5(\varepsilon)\|\hat{Z}\|_{C(0,\infty)}\|f\|_{C(0,\infty)}).$$

So, letting $T \to \infty$, we get $x_\varepsilon \in C(0, \infty)$, since the right-hand part of the latter inequality does not depend on T.

Hence, by Lemma 6.4, a solution y_ε of (3.4) is bounded. Now (3.2) proves the exponential stability. As claimed. □

6.4 Time-Variant NDEs with Discrete Delays

In this section we illustrate Theorem 6.1 in the case of the system

$$\dot{y}_j(t)-\tilde{a}_j\dot{y}_j(t-\tilde{h}_j)+a_jy_j(t-h_j)+\sum_{k=1}^{n} c_{jk}(t)y_k(t-v_{jk}(t)) = 0 \quad (j = 1, \dots, n), \quad (4.1)$$

where $\tilde{h}_j, h_j, a_j$ and $\tilde{a}_j$ are positive constants; $c_{jk}(t)$ $(j, k = 1, \dots, n)$ are continuous functions bounded on $[0, \infty)$ and $v_{jk}(t)$ are continuous functions, satisfying the conditions

$$0 \le v_{jk}(t) \le \eta \quad (j, k = 1, \dots, n;\ t \ge 0),$$

where

$$\eta = \max\{\max_j h_j, \max_j \tilde{h}_j\}.$$

Introduce the matrices

$$A = diag\ (a_j)_{j=1}^n, \tilde{A} = diag\ (\tilde{a}_j)_{j=1}^n \text{ and } C(t) = (c_{jk}(t))_{j,k=1}^n.$$

Simple calculations show that in the considered case we have

$$V_1 = \max_{j=1,\dots,n} \tilde{a}_j \text{ and } V_0 \le \max_{j=1,\dots,n} a_j + \sup_{t\ge 0}\|C(t)\|_n.$$

Theorem 6.2 *Let each of the scalar equations*

$$s = se^{\tilde{h}_j s}\tilde{a}_j + e^{h_j s}a_j \quad (j = 1, \dots, n) \quad (4.2)$$

have a positive root. In addition, let

$$\sum_{k=1}^{n} \sup_{t \geq 0} |c_{jk}(t)| < a_j \quad (j = 1, \ldots, n). \tag{4.3}$$

Then system (4.1) *is exponentially stable.*

This theorem is proved in the next section. Its proof is based on Theorem 6.1.

Note that from the existence of the positive roots to (4.2) it follows that

$$V_1 = \max_{j=1,\ldots,n} \tilde{a}_j < 1.$$

As it was noted in Sect. 5.9, if there are positive numbers λ_j, such that

$$\tilde{a}_j e^{\tilde{h}_j \lambda_j} \lambda_j + a_j e^{h_j \lambda_j} \leq \lambda_j, \tag{4.4}$$

then Eq. (4.2) has a positive root $\psi_j \leq \lambda_j$. In particular, if

$$e^{\tilde{h}_j} \tilde{a}_j + e^{h_j} a_j \leq 1, \tag{4.5}$$

then (4.2) has a positive root $\psi_j \leq 1$.

To investigate the level of conservatism of the sufficient condition of stability given in Theorem 6.2, consider the equation

$$\dot{y}(t) - a\dot{y}(t - \tilde{h}) + by(t - h) = by(t - \hat{h}),$$

where $h, \tilde{h}, \hat{h}, a, b$ are positive constants. This equation is not exponentially stable since its characteristic function

$$z - ze^{\tilde{h}z} a + (e^{hz} - e^{\hat{h}z}) b$$

has a root at $z = 0$. Similarly, considering the system

$$\dot{y}_j(t) - \tilde{a}_j \dot{y}_j(t - \tilde{h}_j) + a_j y(t - h_j) = c_j y(t - \hat{h}_j) \quad (j = 1, \ldots, n),$$

with positive constants $\hat{h}_j$ and c_j, we can assert that its characteristic (diagonal) matrix has a characteristic value at $z = 0$, provided $c_j = a_j$ for at least one index j.

These examples show that condition (4.3) is sharp.

6.5 Proof of Theorem 6.2

Due to Lemma 5.13, the Green function (the fundamental solution) $\hat{Z}_j(t)$ of the equation

$$\dot{y}_j(t) - \tilde{a}_j \dot{y}_j(t - \tilde{h}_j) + a_j y_j(t - h_j) = 0 \quad (j = 1, \ldots, n), \tag{5.1}$$

is positive and satisfies the equality

$$\int_0^\infty \hat{Z}_j(t)dt = \frac{1}{a_j}. \tag{5.2}$$

Consider the nonhomogeneous system

$$\dot{x}_j(t) - a_j\dot{x}_j(t-\tilde{h}_j) + b_j x_j(t-h_j) = \sum_{k=1}^n c_{jk}(t)x_k(t-v_{jk}(t)) + f_j(t) \quad (j=1,\dots,n), \tag{5.3}$$

with scalar continuous functions f_j and the zero initial conditions. By the Variation of Constants formula we can write

$$x_j(t) = \int_0^t \hat{Z}_j(t-s)(F_j(s)+f_j(s))ds,$$

where

$$F_j(t) = \sum_{k=1}^n c_{jk}(t)x_k(t-v_{jk}(t)).$$

For the brevity in the present section put

$$|u|_C = \sup_{t\ge 0}|u(t)|$$

for a scalar continuous function $u(t)$. Hence, by (5.2),

$$|x_j|_C \le \frac{1}{a_j}|F_j|_C + m_j, \text{ where } m_j = \frac{|f_j|_C}{b_j}.$$

Take into account that

$$|F_j|_C \le \sum_{k=1}^n |c_{jk}|_C|x_k|_C$$

and introduce the matrix M whose entries are

$$q_{jk} = \frac{1}{b_j}|c_{jk}|_C \quad (j,k=1,\dots,n).$$

Consequently,

$$column\ (|x_j|_C) \le M\ column(|x_j|_C) + column\ (m_j).$$

Thanks to condition (4.3) we have

$$\sum_{k=1}^{n} q_{jk} < 1 \quad (j = 1, \ldots, n)$$

and therefore, due to Corollary 2.16 the spectral radius of M is less than one. Hence,

$$column\ (|x_j|_C) \leq (I - M)^{-1} column\ (m_j),$$

where I is the unit matrix. So a solution of system (5.3) is bounded. Now the required result is due to Theorem 6.1. □

6.6 The L^p-version of the Generalized Bohl-Perron Principle

Again $\tilde{A}_j(t), A_k(t)$ $(j = 1, \ldots, m_1;\ k = 1, \ldots, m)$ are $n \times n$ variable matrices defined on the positive half-line;

$$A(t, \tau), \tilde{A}(t, \tau) \quad (t \geq 0;\ \tau \in [0, \eta])$$

are $n \times n$ variable matrices; In addition, $A_k(t), A(t, \tau)$ are continuous in t and integrable in τ; $\tilde{A}_j(t), \tilde{A}(t, \tau)$ have continuous derivatives in t. At $u \in L^p(0, \infty)$, operators E_1, E_2 are defined by

$$(E_1 u)(t) = \sum_{k=1}^{m_1} \tilde{A}_k(t) u(t - \tilde{h}_k) + \int_0^{\eta} \tilde{A}(t, s) u(t - s) ds$$

and

$$(E_2 u)(t) = \sum_{k=1}^{m} A_k(t) u(t - h_k(t)) + \int_0^{\eta} A(t, s) u(t - s) ds \quad (t \geq 0),$$

where

$$0 < \tilde{h}_1 < \cdots < \tilde{h}_{m_1} \leq \eta \quad (m_1 < \infty)$$

are constants, $h_j(t)$ $(j = 1, \ldots, m < \infty)$ are real continuously differentiable functions.

Again consider the equation

$$\frac{d}{dt}[y(t) - (E_1 y)(t)] = (E_2 y)(t) \tag{6.1}$$

with the initial condition

$$y(t) = \phi(t) \quad (-\eta \le t \le 0) \tag{6.2}$$

for a given $\phi \in C^1(-\eta, 0)$. We consider also the non-homogeneous equation

$$\frac{d}{dt}[x(t) - (E_1x)(t)] = (E_2x)(t) + f(t) \quad (t \ge 0) \tag{6.3}$$

with a given vector function $f \in L^p(0, \infty)$ and the zero initial condition

$$x(t) \equiv 0 \quad (-\eta \le t \le 0). \tag{6.4}$$

Throughout this section it is assumed that

$$0 \le h_j(t) \le \eta \text{ and } \dot{h}_j(t) < 1 \quad (j = 1, \dots, m;\ t \ge 0). \tag{6.5}$$

In addition,

$$\|A_j(t)\|_n \quad (j = 1, \dots, m),\ \|\tilde{A}_k'(t)\|_n \quad (k = 1, \dots, m_1)$$

are bounded on $[0, \infty)$,

$$\int_0^\eta \sup_{t \ge 0} \; (\|\tilde{A}_t'(t, s)\|_n + \|A(t, s)\|_n) ds < \infty$$

and

$$V_1 := \sum_{k=1}^{m_1} \sup_{t \ge 0} \|\tilde{A}_k(t)\|_n + \int_0^\eta \sup_{t \ge 0} \|\tilde{A}(t, s)\|_n ds < 1. \tag{6.6}$$

Solutions of problems (6.1), (6.2), (6.3) and (6.4) are defined as in Sect. 6.1. The existence and uniqueness of the considered solutions are due to Corollary 6.1 and Lemma 6.3.

Now we are in a position to formulate the main result of the section.

Theorem 6.3 *Let condition* (6.5) *and* (6.6) *hold, and for a* $p \in [1, \infty)$ *and any* $f \in L^p(0, \infty)$ *problem* (6.3), (6.4) *have a solution* $x(t) \in L^p(0, \infty)$. *Then Eq.* (6.1) *is exponentially stable.*

This theorem is proved in the next section.

Remark 6.2 We consider Eq. (6.1) for the sake of the simplicity. In the arguments of the Proof of Theorem 6.3 one can replace the operator E_2 by the operator E defined in Sect. 6.1, if it is bounded on $L^p(0, \infty)$.

6.7 Proof of Theorem 6.3

We need the following simple lemma.

Lemma 6.7 *Let $h(t)$ $(t \geq 0)$ be a differentiable function with the properties*

$$0 < h(t) \leq \eta \text{ and } \dot{h}(t) < 1 \quad (t \geq 0).$$

Then for any $u \in L^p(-\eta, \infty)$ $(1 \leq p < \infty)$ one has

$$\|u(t - h(t))\|_{L^p(0,T)} \leq \sqrt[p]{\chi(h)} \|u\|_{L^p(-h(0),T)} \quad (T > 0),$$

where

$$\chi(h) = \frac{1}{\inf_{t \geq 0} 1 - \dot{h}(t)}.$$

Proof Obviously,

$$\int_0^T \|u(t - h(t))\|_n^p dt = \int_0^T \|u(t - h(t))\|_n^p \frac{1 - \dot{h}(t)}{1 - \dot{h}(t)} dt$$

$$\leq \chi(h) \int_0^T \|u(t - h(t))\|_n^p (1 - \dot{h}(t)) dt = \chi(h) \int_{-h(0)}^{\infty} \|u(s)\|_n^p ds.$$

As claimed. □

Rewrite (6.1) as

$$\dot{y}(t) - (E_1 \dot{y})(t) = (E_1' y)(t) + (E_2 y)(t) \quad (t \geq 0), \tag{7.1}$$

where

$$(E_1' y)(t) = \sum_{k=1}^{m_1} \tilde{A}_k'(t) y(t - \tilde{h}_k) + \int_0^{\eta} \tilde{A}_t'(t, s) y(t - s) ds.$$

Lemma 6.8 *For any $T > 0$ and a $p \geq 1$ one has $\|E_1 u\|_{L^p(0,T)} \leq V_1 \|u\|_{L^p(-\eta,T)}$ and there is a constant V_2 independent on $T > 0$, such that*

$$\|(E_1' + E_2) u\|_{L^p(0,T)} \leq V_2 \|u\|_{L^p(-\eta,T)} \quad (u \in L^p(-\eta, T)).$$

Proof Let $u \in L^\infty(-\eta, T)$. Then

$$\|(E_1u)(t)\|_n \leq \sum_{k=1}^{m_1} \|\tilde{A}_k(t)u(t-\tilde{h}_k)\|_n + \int_0^\eta \|\tilde{A}(t,s)u(t-s)\|_n ds$$
$$\leq \|u\|_{L^\infty(-\eta,T)}\Big(\sum_{k=1}^{m_1} \|\tilde{A}_k(t)\|_n + \int_0^\eta \|\tilde{A}(t,s)\|_n ds\Big)$$
$$\leq V_1\|u\|_{L^\infty(-\eta,T)} \quad (0 \leq t \leq T).$$

So for L^∞ the first inequality is proved. Now let $u \in L^1(-\eta, T)$. Then

$$\|E_1u\|_{L^1(0,T)} \leq \sum_{k=1}^{m_1} \|\tilde{A}_k(t)u(t-\tilde{h}_k)\|_{L^1(0,T)} + \int_0^\eta \|\tilde{A}(t,s)u(t-s)\|_{L^1(0,T)} ds$$
$$\leq \|u\|_{L^1(-\eta,T)}\Big(\sum_{k=1}^{m_1} \sup_t \|\tilde{A}_k(t)\|_n + \int_0^\eta \sup_t \|\tilde{A}(t,s)\|_n ds\Big)$$
$$= V_1\|u\|_{L^1(-\eta,T)}.$$

Hence for all $p \geq 1$ the first inequality is due to the Riesz-Thorin theorem (see Sect. 1.3). Similarly by Lemma 6.7 the second inequality can be proved. □

Lemma 6.9 *If for any* $f \in L^p(0,\infty)$ $(1 \leq p < \infty)$ *a solution of problem* (6.3), (6.4) *and its derivative are in* $L^p(0,\infty)$, *and operators* E_1, E_2 *and* E_1' *are bounded in* $L^p(0,\infty)$, *then any solution of problem* (6.1), (6.2) *and its derivative are in* $L^p(-\eta,\infty)$.

Proof Let $y(t)$ be a solution of problem (6.1), (6.2). For a positive constant $\nu > 0$ put

$$\hat{\phi}_\nu(t) = \begin{cases} e^{-\nu t}\phi(0) & \text{if } t \geq 0, \\ \phi(t) & \text{if } -\eta \leq t < 0 \end{cases}$$

and $x_0(t) = y(t) - \hat{\phi}_\nu(t)$. We can write $d\hat{\phi}_\nu(t)/dt = -\nu e^{-\nu t}\phi(0) \quad (t \geq 0)$ and

$$\frac{d}{dt}[x_0(t) - (E_1x_0)(t)] = (E_2x_0)(t) + f_\nu(t) \quad (t > 0),$$

where

$$f_\nu(t) = -\frac{d\hat{\phi}_\nu(t)}{dt} + \frac{d(E_1\hat{\phi}_\nu)(t)}{dt} + (E_2\hat{\phi}_\nu)(t)$$
$$= -\frac{d\hat{\phi}_\nu(t)}{dt} + \Big(E_1\frac{d\hat{\phi}_\nu}{dt}\Big)(t) + (E_1'\hat{\phi}_\nu)(t) + (E_2\hat{\phi}_\nu)(t).$$

Besides, (6.4) holds with $x(t) = x_0(t)$ and $f(t) = f_\nu(t)$. Since

$$\hat{\phi}_\nu \in L^p(-\eta, \infty) \text{ and } \frac{d\hat{\phi}_\nu(t)}{dt} \in L^p(-\eta, \infty),$$

by the previous lemma we have $f_\nu \in L^p(-\eta, \infty)$. Due to the hypothesis of this lemma, $x_0, \dot{x}_0 \in L^p(0, \infty)$. Thus $y = x_0 + \hat{\phi}_\nu \in L^p(-\eta, \infty)$, $\dot{y} \in L^p(-\eta, \infty)$. As claimed. □

Lemma 6.10 *Let condition* (6.6) *hold. Then for any solution* $x(t)$ *of problem* (6.3), (6.4) *and all* $T > 0$, *one has*

$$\|\dot{x}\|_{L^p(0,T)} \leq (1 - V_1)^{-1}(V_2\|x\|_{L^p(0,T)} + \|f\|_{L^p(0,T)}) \quad (p \geq 1).$$

Proof By Lemma 6.8 from (6.3) we have

$$\|\dot{x}\|_{L^p(0,T)} \leq V_1\|\dot{x}\|_{L^p(0,T)} + V_2\|x\|_{L^p(0,T)} + \|f\|_{L^p(0,T)}.$$

Hence the condition $V_1 < 1$ implies the required result. □

Lemma 6.11 *Any function* $\xi \in L^p(0, \infty)$ $(1 \leq p < \infty)$ *with* $\dot{\xi} \in L^p(0, \infty)$ *is bounded on* $[0, \infty)$. *Moreover,*

$$\|\xi\|^p_{C(0,\infty)} \leq p\|\xi\|^{p-1}_{L^p(0,\infty)}\|\dot{\xi}\|_{L^p(0,\infty)} \quad \text{if } 1 < p < \infty,$$

and

$$\|\xi\|_{C(0,\infty)} \leq \|\dot{\xi}\|_{L^1(0,\infty)}.$$

Proof For simplicity, in this proof put $\|\xi(t)\|_n = |\xi(t)|$. First consider the case $p = 1$. Since,

$$\begin{aligned}\frac{d|\xi(t)|}{dt} &= \lim_{h\to 0}\frac{|\xi(t+h)| - |\xi(t)|}{h}\\ &\leq \lim_{h\to 0}\frac{|\xi(t+h) - \xi(t)|}{h} = |\dot{\xi}(t)|,\end{aligned}$$

we obtain

$$|\xi(t)| = -\int_t^\infty \frac{d|\xi(t_1)|}{dt_1}dt_1 \leq \int_t^\infty |\dot{\xi}(t_1)|dt_1 \leq \|\dot{\xi}\|_{L^1(0,\infty)} \quad (t \geq 0).$$

Assume that $1 < p < \infty$. Then by the Hólder inequality

$$|\xi(t)|^p = -\int_t^\infty \frac{d|\xi(t_1)|^p}{dt_1}dt_1 = -p\int_t^\infty |\xi(t_1)|^{p-1}\frac{d|\xi(t_1)|}{dt_1}dt_1$$
$$\leq p\int_t^\infty |\xi(t)|^{p-1}|\dot{\xi}(t_1)|dt_1$$
$$\leq p[\int_t^\infty |\xi(t_1)|^{q(p-1)}dt_1]^{1/q}[\int_t^\infty |\dot{\xi}(t_1)|^p dt_1]^{1/p}$$

where $q = p/(p-1)$. Since $q(p-1) = p$, we get the required inequality. □

Proof of Theorem 6.3 Substituting

$$y(t) = y_\varepsilon(t)e^{-\varepsilon t} \tag{7.2}$$

with an $\varepsilon > 0$ into (6.1), we obtain the equation

$$\dot{y}_\varepsilon - \varepsilon y_\varepsilon - E_{\varepsilon,1}\dot{y}_\varepsilon + \varepsilon E_{\varepsilon,1}y_\varepsilon = (E'_{\varepsilon,1} + E_{\varepsilon,0})y_\varepsilon, \tag{7.3}$$

where

$$(E_{\varepsilon,1}f)(t) = \sum_{k=1}^{m_1} e^{\tilde{h}_k\varepsilon}\tilde{A}_k(t)f(t-\tilde{h}_k) + \int_0^\eta e^{s\varepsilon}\tilde{A}(t,s)f(t-s)ds,$$
$$(E_{\varepsilon,0}f)(t) = \sum_{k=1}^{m} A_k(t)e^{h_k(t)\varepsilon}f(t-h_k(t)) + \int_0^\eta A(t,s)e^{s\varepsilon}f(t-s)ds$$

and

$$(E'_{\varepsilon,1}f)(t) = \sum_{k=1}^{m_1} e^{\tilde{h}_k\varepsilon}\tilde{A}'_k(t)f(t-\tilde{h}_k) + \int_0^\eta e^{s\varepsilon}\tilde{A}'_t(t,s)f(t-s)ds.$$

Rewrite (7.3) as

$$\frac{d}{dt}[y_\varepsilon - E_{\varepsilon,1}y_\varepsilon] = W_\varepsilon y_\varepsilon + E_{\varepsilon,0}y_\varepsilon, \tag{7.4}$$

where $W_\varepsilon := \varepsilon I - \varepsilon E_{\varepsilon,1}$. Furthermore, introduce in $L^p(0,\infty)$ the operator $\hat{Z} : f \to x$ where $x(t)$ is the solution of problem (7.3), (7.4). That is, $\hat{Z}$ solves problem (7.3), (7.4).

By the hypothesis of the theorem, we have $x = \hat{Z}f \in L^p(0,\infty)$ for any $f \in L^p(0,\infty)$. So $\hat{Z}$ is defined on the whole space $L^p(0,\infty)$. It is closed, since the uniqueness of solutions to problem (7.3), (7.4) is due to Corollary 6.1. Therefore $\hat{Z} : L^p(0,\infty) \to L^p(0,\infty)$ is bounded according to the Closed Graph Theorem (see Sect. 1.3). So the norm $\|\hat{Z}\|_{L^p(0,\infty)}$ is finite. Consider now the equation

$$\frac{d}{dt}[x_\varepsilon - E_{\varepsilon,1}x_\varepsilon] = W_\varepsilon x_\varepsilon + E_{\varepsilon,0}x_\varepsilon + f \tag{7.5}$$

with the zero initial conditions. Subtract (7.3) from (7.5), with $w(t) = x_\varepsilon(t) - x(t)$, where x and x_ε are solutions of problems (7.3), (7.4), (7.5) and (7.4), respectively. Then

$$\frac{d}{dt}[w - E_1 w] = E_2 w + F_\varepsilon \tag{7.6}$$

where

$$F_\varepsilon = W_\varepsilon x_\varepsilon + (E_{\varepsilon,0} - E_2)x_\varepsilon + \frac{d}{dt}(E_{\varepsilon,1} - E_1)x_\varepsilon. \tag{7.7}$$

It is simple to check that $W_\varepsilon \to 0$, $E'_{\varepsilon,1} \to E'_1$ and $E_{\varepsilon,1} \to E_1$ in the operator norm of $L^p(0,\infty)$ as $\varepsilon \to 0$.

For the brevity in this proof put $\|.\|_{L^p(0,T)} = |.|_T$ for a finite $T > 0$. We have

$$\frac{d}{dt}(E_{\varepsilon,1} - E_1)x_\varepsilon = (E_{\varepsilon,1} - E_1)\dot{x}_\varepsilon + (E'_{\varepsilon,1} - E'_1)x_\varepsilon.$$

So

$$\left|\frac{d}{dt}(E_{\varepsilon,1} - E_1)x_\varepsilon\right|_T \le a_1(\varepsilon)|\dot{x}_\varepsilon|_T + a_2(\varepsilon)|x_\varepsilon|_T, \tag{7.8}$$

where $a_j(\varepsilon) \to 0$ as $\varepsilon \to 0$ for an index j.

But according to Lemma 6.8, for a sufficiently small ε, we have $|E_{\varepsilon,1}x_\varepsilon|_T \le e^{\varepsilon\eta}V_1|x_\varepsilon|_T$ with

$$e^{\varepsilon\eta}V_1 < 1. \tag{7.9}$$

Due to Lemma 6.10, from (7.5), the inequality

$$|\dot{x}_\varepsilon|_T \le (1 - e^{\varepsilon\eta}V_1)^{-1}(|E'_{\varepsilon,1} + W_\varepsilon|_T|x_\varepsilon|_T + |f|_T)$$

follows. Since $W_\varepsilon \to 0$, $E'_{\varepsilon,1} \to E'_1$, the norm $|E'_{\varepsilon,1} + W_\varepsilon|_T$ is bounded uniformly with respect to ε and $T > 0$. So for a sufficiently small $\varepsilon_0 > 0$, there is a constant c_1, such that

$$|\dot{x}_\varepsilon|_T \le c_1(|x_\varepsilon|_T + |f|_T) \quad (\varepsilon < \varepsilon_0;\ T > 0).$$

Now (7.8) implies

$$|\frac{d}{dt}(E_{\varepsilon,1} - E_1)x_\varepsilon|_T \le a_1(\varepsilon)c_1(|x_\varepsilon|_T + |f|_T) + a_2(\varepsilon)|x_\varepsilon|_T \le a_3(\varepsilon)(|x_\varepsilon|_T + |f|_T). \tag{7.10}$$

Furthermore,

$$|(E_{\varepsilon,0} - E_2)x_\varepsilon|_T \le a_4(\varepsilon)|x_\varepsilon|_T.$$

Thus from (7.10) we deduce that

$$|F_\varepsilon|_T \le a_5(\varepsilon)(|x_\varepsilon|_T + |f|_T)$$

uniformly in $T > 0$. By (7.6) $x_\varepsilon - x = \hat{Z}F_\varepsilon$. So

$$|x_\varepsilon - x|_T \le \|\hat{Z}\|_{L^p(0,\infty)}a_5(\varepsilon)(|x_\varepsilon|_T + |f|_T).$$

For a sufficiently small ε, we have $q(\varepsilon) := \|\hat{Z}\|_{L^p(0,\infty)}a_5(\varepsilon) < 1$. Thus

$$|x_\varepsilon|_T \le (1 - q(\varepsilon))^{-1}(|x|_T + \|\hat{Z}\|_{L^p(0,\infty)}a_5(\varepsilon)|f|_T).$$

By the hypothesis of this theorem, $x \in L^p(0, \infty)$. This gives us the inequality

$$|x_\varepsilon|_T \le (1 - q(\varepsilon))^{-1}(\|x\|_{L^p(0,\infty)} + a_5(\varepsilon)\|\hat{Z}\|_{L^p(0,\infty)}\|f\|_{L^p(0,\infty)}).$$

So, letting $T \to \infty$, we get $x_\varepsilon \in L^p(0, \infty)$. Hence, Lemma 6.10 implies $\dot{x}_\varepsilon \in L^p(0, \infty)$.

Consequently, due Lemma 6.9, a solution y_ε of (6.3) and its derivative are in $L^p(0, \infty)$. Making use Lemma 6.11 we can assert that a solution of (7.3) is bounded. Now (7.2) proves the exponential stability. As claimed. □

6.8 Time-Variant NDEs Close to Autonomous Systems

Let us consider the equation

$$\dot{x}(t) - \tilde{A}\int_0^\eta \dot{x}(t-s)d\tilde{\mu}(s) + C\int_0^\eta x(t-s)d\mu(s) = E_2x(t) + f(t) \quad (t \ge 0), \tag{8.1}$$

where as above, $\tilde{A}$ and C are constant $n \times n-$matrices, and μ, $\tilde{\mu}$ are scalar nondecreasing functions with finite numbers of jumps; E_2 is again defined by

$$(E_2u)(t) = \sum_{k=1}^{m} A_k(t)u(t - h_k(t)) + \int_0^{\eta} A(t,s)u(t-s)ds,$$

where $h_j(t)$ are real differentiable functions, such that

$$0 \le h_j(t) \le \eta \text{ and } \dot{h}_j(t) < 1 \quad (t \ge 0;\ j = 1, \dots, m);$$

$A_k(t)$ $(; k = 1, \dots, m)$ are $n \times n$ continuous matrix functions defined and bounded on the positive half-line; $A(t, \tau)$ $(t \ge 0;\ \tau \in [0, \eta])$ is an $n \times n$ variable matrix, integrable in τ, and continuous and uniformly bounded in $t \ge 0$.

Without loss of generality suppose that

$$var(\mu) = var(\tilde{\mu}) = 1.$$

In addition, assume that

$$\|\tilde{A}\|_n < 1. \tag{8.2}$$

Furthermore, due to Lemma 6.7

$$\|E_2\|_{L^2(-\eta,\infty)\to L^2(0,\infty)} \le q(E_2) := \sum_{k=1}^{m} \frac{\sup_{t\ge0} \|A_k(t)\|_n}{\inf_{t\ge0} \sqrt{1 - \dot{h}_k(t)}} + \int_0^{\eta} \sup_{t\ge0} \|A(t,s)\|_n ds. \tag{8.3}$$

Certainly Eq. (8.1) can be written in the form (6.3) with

$$-C \int_0^{\eta} x(t-s)d\mu(s) + E_2x(t)$$

instead of $E_2x(t)$. Besides $\tilde{h}_k$ are the points of the jumps of $\tilde{\mu}$,

$$\tilde{A}_k(t) = (\tilde{\mu}(h_k + 0) - \tilde{\mu}(h_k - 0))\tilde{A},$$
$$\tilde{A}_k(t,s) \equiv \tilde{A}_k(s) = \frac{d\tilde{\mu}(s)}{ds}\tilde{A},$$

etc.

Furthermore, put

$$Q(z) = zI - z\tilde{A} \int_0^{\eta} e^{-zs} d\tilde{\mu}(s) + C \int_0^{\eta} e^{-zs} d\mu(s).$$

Assume that all the characteristic values of $Q(z)$ are in the open left half plane C_-. Then

$$Z_0(t) = \frac{1}{2\pi} \int_{-\infty}^{\infty} e^{i\omega t} Q^{-1}(i\omega) d\omega$$

is the fundamental solution of the autonomous equation

$$\dot{z}(t) - \tilde{A} \int_0^{\eta} \dot{z}(t-s) d\tilde{\mu}(s) + C \int_0^{\eta} z(t-s) d\mu(s) = 0 \quad (t \geq 0). \tag{8.4}$$

Furthermore, use the operator $\hat{Z}_0$ defined on $L^2(0,\infty)$ by

$$\hat{Z}_0 f(t) = \int_0^t Z_0(t-t_1) f(t_1) dt_1 \quad (f \in L^2(0,\infty)).$$

Then due to (8.1) we get $x = \hat{Z}_0(E_2 x + f)$. Assume that

$$q(E_2)\ \|\hat{Z}_0\|_{L^2(0,\infty)} < 1. \tag{8.5}$$

Then

$$\|x\|_{L^2(0,\infty)} \leq \frac{\|\hat{Z}_0 f\|_{L^2(0,\infty)}}{1 - q(E_2)\|\hat{Z}_0\|_{L^2(0,\infty)}}.$$

Now Theorem 6.3 implies

Lemma 6.12 *Let conditions* (8.2) *and* (8.5) *hold. Then the equation*

$$\dot{y}(t) - \tilde{A} \int_0^{\eta} \dot{y}(t-s) d\tilde{\mu}(s) + \int_0^{\eta} Cy(t-s) d\mu(s) = E_2 y(t) \quad (t \geq 0), \tag{8.6}$$

is exponentially stable.

Put

$$\zeta_0 = \frac{2\|C\|_n}{1 - \|\tilde{A}\|_n}$$

and

$$\theta(Q) = \sup_{-\zeta_0 \leq \omega \leq \zeta_0} \|Q^{-1}(i\omega)\|_n.$$

Then due to Theorem 5.1

$$\|\hat{Z}_0\|_{L^2(0,\infty)} < \theta(Q). \tag{8.7}$$

Thus condition (8.5) is provided by the inequality

$$q(E_2)\theta(Q) < 1. \tag{8.8}$$

Recall that

$$g(A) = (N_2^2(A) - \sum_{k=1}^{n} |\lambda_k(A)|^2)^{1/2}$$

for an $n \times n$ matrix A. Put

$$B(z) = \tilde{A} \int_0^{\eta} z\ exp(-zs)d\tilde{\mu}(s) - C \int_0^{\eta} exp(-zs)d\mu(s).$$

So $Q(z) = zI - B(z)$. As it was shown in Sect. 5.4, the inequality

$$\|[Q(z)]^{-1}\|_n \le \Gamma(Q(z)) \ (z \in \mathbf{C}) \tag{8.9}$$

is valid, where

$$\Gamma(Q(z)) = \sum_{k=0}^{n-1} \frac{g^k(B(z))}{\sqrt{k!}\rho^{k+1}(Q(z))}$$

and $\rho(Q(z))$ is the smallest absolute value of the eigenvalues of $Q(z)$ for a fixed z:

$$\rho(Q(z)) = \min_{k=1,\dots,n} |\lambda_k(Q(z))|.$$

Thus

$$\|\hat{Z}_0\|_{L^2(0,\infty)} \le \theta(Q) \le \Gamma_0(Q), \tag{8.10}$$

where

$$\Gamma_0(Q) = \sup_{-\zeta_0 \le \omega \le \zeta_0} \Gamma(Q(i\omega))$$

(see also Lemma 5.7). Hence, due to Lemma 6.12 we arrive at the following result.

Theorem 6.4 *Let all the zeros of Q be in C_- and the conditions* (8.2) *and* $q(E_2)\Gamma_0(Q) < 1$ *hold. Then Eq.* (8.6) *is exponentially stable.*

Denote

$$\hat{g}(B) := \sup_{\omega \in [-\zeta_0, \zeta_0]} g(B(i\omega)) \text{ and } \hat{\rho}(Q) := \inf_{\omega \in [-\zeta_0, \zeta_0]} \rho(Q(i\omega)).$$

Then we have

$$\Gamma_0(Q) \le \hat{\Gamma}(Q), \text{ where } \hat{\Gamma}(Q) := \sum_{k=0}^{n-1} \frac{\hat{g}^k(B)}{\sqrt{k!}\hat{\rho}^{k+1}(Q)}.$$

Now Theorem 6.4 implies

Corollary 6.2 *Let all the zeros of det $Q(z)$ be in C_- and the conditions* (8.2), *and* $q(E_2)\hat{\Gamma}(Q) < 1$ *hold. Then Eq.* (8.6) *is exponentially stable.*

Thanks to the definition of $g(A)$, for all $\omega \in \mathbf{R}$ one can write

$$g(B(i\omega)) \le N_2(B(i\omega)) \le |\omega| N_2(\tilde{A}) + N_2(C).$$

That inequality can be improved, for example in the case

$$Q(z) = zI - z\tilde{A}e^{-z\tilde{h}} + C\int_0^\eta e^{-zs}d\mu(s). \tag{8.11}$$

Indeed, due to the property $g^2(M) \le N_2^2(M - M^*)/2$ for an arbitrary matrix M (see Sect. 2.1) we can write

$$g(B(i\omega)) = g(ie^{i\omega\tilde{h}}B(i\omega)) \le \frac{1}{\sqrt{2}}[\, |\omega| N_2(\tilde{A} - \tilde{A}^*)$$

$$+ N_2(\int_0^\eta e^{-i\omega(s-\tilde{h})}d\mu(s)C + \int_0^\eta e^{i\omega(s-\tilde{h})}d\mu(s)C^*)].$$

So in the case (8.11) we get

$$\hat{g}(Q) \le \frac{\zeta_0}{\sqrt{2}} N_2(\tilde{A} - \tilde{A}^*) + \sqrt{2}N_2(C).$$

In the rest of this section we suppose that $\tilde{A}$ and C commute. So the eigenvalues of $Q(z)$ for a fixed z can be written as

$$\lambda_j(Q(z)) = z - z\int_0^\eta e^{-zs}d\tilde{\mu}(s)\lambda_j(\tilde{A}) + \int_0^\eta e^{-zs}d\mu(s)\lambda_j(C),$$

Recall that for two commuting matrices A_1 and A_2 we have $g(A_1 + A_2) \le g(A_1) + g(A_2)$ (see Sect. 2.1). Then

$$g(B(i\omega)) \le g(i\omega\tilde{A}) + g(C) = |\omega| g(\tilde{A}) + g(C).$$

Furthermore, suppose $\lambda_k(C)$ and $\lambda_k(\tilde{A})$ $(k = 1, \dots, n)$ are positive and put

$$\nu_k = \frac{2\lambda_k(C)}{1 - \lambda_k(\tilde{A})}.$$

If, in addition,

$$\eta\nu_k < \pi/2 \tag{8.12a}$$

and

$$d_k(\mu, \tilde{\mu}) := \lambda_k(C) \int_0^\eta \cos(\tau\nu_k) d\mu - \nu_k\lambda_k(\tilde{A}) \int_0^\eta \sin(\tau\nu_k) d\tilde{\mu} > 0 \quad (k = 1, \dots, n), \tag{8.12b}$$

then by Lemma 5.12 all the characteristic values of Q are in C_- and

$$\inf_{\omega\in\mathbf{R}} |\lambda_j(Q(i\omega))| \geq \tilde{d}_{com} := \min_k d_k(\mu, \tilde{\mu}) \quad (j = 1, \dots, n). \tag{8.13}$$

So

$$\hat{\Gamma}(Q) \leq \Gamma_{com}(Q) := \sum_{k=0}^{n-1} \frac{g^k(C, \tilde{A})}{\sqrt{k!}\tilde{d}_{com}^{k+1}}.$$

Now Corollary 6.2 implies

Corollary 6.3 *Let $\tilde{A}$ and C be commuting matrices with positive eigenvalues. Let the conditions* (8.2), (8.12a), (8.12b) *and $q(E_2)\Gamma_{com}(Q) < 1$ be fulfilled. Then Eq.* (8.6) *is exponentially stable.*

Example Consider the system

$$\dot{y}_j(t) - \tilde{a}\dot{y}_j(t - \tilde{h}) + \sum_{k=1}^{2} c_{jk}y_k(t) = \sum_{k=1}^{2} a_{jk}(t)y_k(t - h(t)) \quad (j = 1, 2;\ t \geq 0), \tag{8.14}$$

where

$$0 < \tilde{a} < 1,\ 0 \leq h(t) \leq \tilde{h} \text{ and } \dot{h}(t) < 1;$$

c_{jk} are real constants, and $A(t) = (a_{jk}(t))_{j,k=1}^2$. So $\tilde{A} = \tilde{a}I$,

$$Q(z) = z(1 - \tilde{a}e^{-z\tilde{h}})I + C$$

with $C = (c_{jk})$, $g(\tilde{A}) = 0$ and

$$g(C, \tilde{A}) = g(C) = |c_{12} - c_{21}|.$$

The eigenvalues of $Q(z)$ for a fixed z are

$$\lambda_j(Q(z)) = z - z\,\tilde{a}e^{-z\tilde{h}} + \lambda_j(C).$$

Suppose $\lambda_k(C)$ $(k = 1, 2)$ are positive and put

$$v_k = \frac{2\lambda_k(C)}{1 - \tilde{a}}.$$

If

$$\tilde{h}v_k < \pi/2 \text{ and } d_k = \lambda_k(C) - v_k\tilde{a}\sin(\tilde{h}v_k) > 0 \quad (k = 1, 2), \tag{8.15}$$

then by (8.13), the characteristic values of Q are in C_-,

$$\inf_{\omega\in\mathbf{R}} |\lambda_k(Q(i\omega))| \geq \hat{d} := \min_{k=1,2} d_k$$

and

$$\hat{\Gamma}(Q) \leq \Gamma_1 := \frac{1}{\hat{d}}\left(1 + \frac{g(C)}{\hat{d}}\right).$$

In the considered example $E_2y(t) = A(t)y(t - h(t))$ and

$$q(E_2) = \frac{\sup_t \|A(t)\|_n}{\inf_t \sqrt{1 - \dot{h}(t)}}.$$

Thanks to Corollary 6.2 we can assert that system (8.14) is exponentially stable provided the conditions (8.15) and $q(E_2)\Gamma_1 < 1$ hold.

6.9 NDEs with Small Norms of Principal Operators

In this section we derive norm estimates for the Cauchy operators of NDEs in terms of the Cauchy operators of certain differential delay and difference equations. These estimates give us a possibility to apply the results derived in Chaps. 3 and 4 to the stability analysis of NDEs.

Again put

$$Su(t) = \int_0^{\min\{\eta,t\}} d_\tau \tilde{R}(t,\tau)u(t-\tau) \quad (t \geq 0) \text{ and } Su(t) = 0 \quad (t < 0)$$

and

$$(Eu)(t) = \int_0^\eta d_\tau R(s)u(t-s)$$

($u \in L^p(-\eta, T), p \geq 1$). Let $\hat{G}_R$ be the Cauchy operator to the differential delay equation

$$\dot{w}(t) - \int_0^\eta d_\tau R(t,\tau)w(t-\tau) = 0 \quad (t > 0). \tag{9.1}$$

That is,

$$\hat{G}_R f(t) = \int_0^t G_R(t,s)f(s)ds \quad (t \geq 0),$$

where $G_R(t,s)$ is the fundamental solution to Eq. (9.1).

As it is shown in Sect. 1.9, under condition (1.3a), for all $p \geq 1$ we have

$$V_{L^p}(R) := \|E\|_{L^p(-\eta,\infty)\to L^p(0,\infty)} < \infty \text{ and } V_C(R) := \|E\|_{C(-\eta,\infty)\to C(0,\infty)} < \infty. \tag{9.2}$$

Besides, $V_{L^2}(R) \leq V(R)$, where $V(R) = \|(v_{jk})_{j,k=1}^n\|_n$, and

$$V_{L^1}(R) \leq \sum_{j=1}^n \sqrt{\sum_{k=1}^n v_{jk}^2}.$$

In the concrete cases these inequalities can be simplified, see for instance Lemmas 6.4 and 6.8. It is clear that

$$\|S\|_{L^p(-\eta,\infty)\to L^p(0,\infty)} \leq V_{L^p}(\tilde{R}) \quad (p \geq 1).$$

Now we are in a position to prove the main result of this section.

Theorem 6.5 *Assume that conditions* (1.3a), (1.3b) *hold and for a* $p \geq 1$, *the inequality*

$$\xi_{L^p} := \|(I-S)^{-1}\hat{G}_R E S\|_{L^p(0,\infty)} < 1$$

is fulfilled, then the Cauchy operator $\hat{Z}$ *to* (1.1) *is bounded in* $L^p(0,\infty)$. *Moreover,*

$$\|\hat{Z}\|_{L^p(0,\infty)} \leq \frac{\|(I-S)^{-1}\hat{G}_R\|_{L^p(0,\infty)}}{1-\xi_{L^p}}. \tag{9.3}$$

Proof Rewrite (1.1) as

$$(\frac{d}{dt} - E)(I - S)x - ESx = f.$$

Since $(I-S)^{-1}$ is the inverse to $I-S$, and $\hat{G}_R$ is the left inverse to $\frac{d}{dt} - E$, we obtain

$$x = (I - S)^{-1}\hat{G}_R(ESx + f).$$

According to the hypothesis of the theorem, the norm of $(I - S)^{-1}\hat{G}_R ES$ is less than one. So $I - (I - S)^{-1}\hat{G}_R ES$ is invertible and

$$x = (I - (I - S)^{-1}\hat{G}_R ES)^{-1}(I - S)^{-1}\hat{G}_R f.$$

This proves the required result. □

Remark 6.3 Since solutions of NDEs are continuous, in our arguments one can replace $L^\infty(0, \infty)$ by $C(0, \infty)$.

Suppose that

$$(V_{L^p}(R)\|\hat{G}_R\|_{L^p(0,\infty)} + 1)V_{L^p}(\tilde{R}) < 1. \tag{9.4}$$

Then $V_{L^p}(\tilde{R}) < 1$ and

$$\|(I - S)^{-1}\|_{L^p(0,\infty)} \le \frac{1}{1 - V_{L^p}(\tilde{R})}.$$

Now (9.4) implies

$$\xi_{L^p} \le \frac{\|\hat{G}_R\|_{L^p} V_{L^p}(R) V_{L^p}(\tilde{R})}{1 - V_{L^p}(\tilde{R})} < 1.$$

Due to the previous theorem we thus we arrive at

Corollary 6.4 *Assume that conditions* (1.3a), (1.3b) *and* (9.4) *hold for a* $p \ge 1$. *Then the Cauchy operator to* (1.1) *is bounded in* $L^p(0, \infty)$. *Moreover,*

$$\|\hat{Z}\|_{L^p(0,\infty)} \le \frac{\|\hat{G}_R\|_{L^p}}{1 - V_{L^p}(\tilde{R})(V_{L^p}(R)\|\hat{G}_R\|_{L^p} + 1)}.$$

Consider the equation

$$\frac{d}{dt}\left[\sum_{k=1}^{m_1}\tilde{A}_k(t)y(t - \tilde{h}_k) + \int_0^\eta \tilde{A}(t, s)y(t - s)ds\right] = (Ey)(t) \quad (t > 0), \tag{9.5}$$

where $\tilde{A}_k(t), \tilde{A}(t,s)$ and $\tilde{h}_k$ are the same as in Sect. 6.2. In the considered case, $S = S_1$, where $S_1 = S_{10} + S_{11} + \cdots + S_{1m_1}$, where

$$S_{10}u(t) = \int_0^{\min\{\eta,t\}} \tilde{A}(t,s)u(t-s)ds \quad (t \geq 0) \text{ and } S_{10}u(t) = 0 \quad (t < 0);$$

$$S_{1k}u(t) = \tilde{A}_k(t)u(t-\tilde{h}_k) \quad (t > \tilde{h}_k) \text{ and } S_{1k}u(t) = 0 \quad (t \leq \tilde{h}_k; k = 1, \ldots, m_1).$$

Taking into account Remarks 6.1 and 6.2, due to Theorems 6.3 we get our next result.

Corollary 6.5 *Assume that the conditions* (1.3a), (1.3b), (9.4) *and*

$$\|\hat{G}_R(I - S_1)^{-1}S_1E\|_{L^p(0,\infty)} < 1$$

hold for a $p \geq 1$. *Then Eq.* (9.5) *is exponentially stable. Besides,* $L^\infty(0,\infty)$ *can be replaced for* $C(0,\infty)$.

For Eq. (9.5), due to Lemmas 6.4 and 6.8, one has $V_{L^p}(\tilde{E}) \leq V_1$ $(p \geq 1)$, where

$$V_1 := \sum_{k=1}^{m_1} \sup_{t \geq 0} \|\tilde{A}_k(t)\|_n + \int_0^\eta \sup_{t \geq 0} \|\tilde{A}(t,s)\|_n ds. \tag{9.6}$$

Now the previous corollary implies

Corollary 6.6 *Assume that the conditions* (1.3a) *and*

$$(V_{L^p}(R)\|\hat{G}_R\|_{L^p(0,\infty)} + 1)V_1 < 1 \quad (p \geq 1)$$

hold for a $p \geq 1$. *Then Eq.* (9.5) *is exponentially stable. Besides,* $L^\infty(0,\infty)$ *can be replaced by* $C(0,\infty)$.

6.10 Stability Conditions Independent of Delay

6.10.1 The General Case

In this section we investigate the following equation in $\mathbf{C}^n$:

$$\frac{d}{dt}(y(t) - E_1y) = A(t)y(t) + Ey(t) \quad (t > 0), \tag{10.1}$$

where

$$E_1 y(t) = \sum_{k=1}^{m_1} \tilde{A}_k(t) y(t - \tilde{h}_k) + \int_0^{\eta} \tilde{A}(t,s) y(t-s) ds$$

and

$$E y(t) = \int_0^{\eta} d_\tau R(t,\tau) y(t-\tau).$$

Matrix functions $\tilde{A}_k(t)$, $\tilde{A}(t,s)$ and $R(t,s)$ are the same as in Sects. 6.2 and 6.9. In addition, $A(t)$ is a continuous matrix valued function bounded on the positive half-line:

$$\|A(.)\|_C := \sup_{t \geq 0} \|A(t)\|_n < \infty.$$

Again denote by $U(t,s)$ $(t \geq s \geq 0)$ the evolution operator of the equation

$$\dot{y} = A(t) y \ (t > 0). \tag{10.2}$$

The following result is due to Lemma 4.23. Let

$$\nu_1 := \sup_{s \geq 0} \int_s^{\infty} \|U(t,s)\|_n dt < \frac{1}{V_{L^1}(R)}. \tag{10.3}$$

Then the Cauchy operator $\hat{G}_1$ of the differential delay equation

$$\dot{x}(t) = A(t)x(t) + Ex(t) \tag{10.4}$$

is subject to the inequality

$$\|\hat{G}_1\|_{L^1(0,\infty)} \leq \chi_1, \text{ where } \chi_1 := \frac{\nu_1}{1 - \nu_1 V_{L^1}(R)}.$$

The latter inequality and Corollary 6.6 with $E + A(t)$ instead of E imply

Corollary 6.7 *Let $V_{L^1}(R)$ be defined by (9.2) and V_1 be defined by (9.6). Assume that the conditions (1.3a) and*

$$((V_{L^1}(R) + \|A(.)\|_C)\nu_1 + 1)V_1 < 1$$

hold. Then the neutral type Eq. (10.1) is exponentially stable. Moreover the Cauchy operator to that equation satisfies the inequality

$$\|\hat{Z}\|_{L^1(0,\infty)} \le \frac{\chi_1}{1-((V_{L^1}(R)+\|A(.)\|_C)\chi_1+1)V_1}.$$

Now consider Eq. (10.1) in space $C(0,\infty)$. Due to Lemma 4.24, if

$$\nu_\infty := \sup_{t\ge 0}\int_0^t \|U(t,s)\|_n ds < \frac{1}{V_C(R)},$$

where $V_C(R) = V_{L^\infty}(R)$ is defined by (9.2), then the Cauchy operator $\hat{G}_1$ of the differential delay Eq. (10.4) satisfies the inequality

$$\|\hat{G}_1\|_C \le \chi_\infty := \frac{\nu_\infty}{1-\nu_\infty V_C(R)}.$$

Now Corollary 6.6 with $E + A(t)$ instead of E implies our next result.

Corollary 6.8 *Assume that the conditions* (1.3a) *and*

$$((V_C(R)+\|A(.)\|_C)\chi_\infty+1)V_1 < 1$$

hold. Then the neutral type Eq. (10.1) *is exponentially stable. More the Cauchy operator to the neutral type Eq.* (10.1) *satisfies the inequality*

$$\|\hat{Z}\|_{C(0,\infty)} \le \frac{\chi_\infty}{1-V_1((V_C(R)+\|A(.)\|_C)\chi_\infty+1)}.$$

6.10.2 Regular Parts with Slowly Varying Coefficients

The results of the present subsection are based on a combined usage of the results of Sect. 6.9 with the freezing method for the differential delay equations.

To formulate and prove the main result of this subsection we need some preliminary results.

Let us consider Eq. (10.1) with the condition

$$\|A(t)-A(s)\|_n \le q_0|t-s| \quad (t,s\ge 0). \tag{10.5}$$

Lemma 6.13 *Let the conditions* (10.5),

$$\epsilon_A := \sup_{s\ge 0}\int_0^\infty \|e^{A(s)t}\|_n dt < \frac{1}{V_C(R)} \tag{10.6}$$

and

$$\hat{\chi}_0 := \sup_{s \geq 0} \int_0^\infty t \| e^{A(s)t} \|_n dt < \frac{1 - \varepsilon_A V_C(R)}{q_0} \tag{10.7}$$

hold. Then the Cauchy operator $\hat{G}_1$ *to the differential delay Eq.* (10.4) *satisfies the inequality* $\|\hat{G}_1\|_{C(0,\infty)} \leq \tau(A, E)$, *where*

$$\tau(A, E) := \frac{\varepsilon_A}{1 - q_0 \hat{\chi}_0 - V_C(R)\varepsilon_A}.$$

To prove this result put

$$\vartheta_U := \sup_{f \in C(0,\infty)} \frac{1}{\|f\|_{C(0,\infty)}} \sup_{t \geq 0} \left\| \int_0^t U(t,s) f(s) ds \right\|_n .$$

The Proof of Lemma 6.13 is based on the following technical lemma.

Lemma 6.14 *Let the condition*

$$\vartheta_U V_C(R) < 1 \tag{10.8}$$

hold. Then the Cauchy operator $\hat{Z}_1$ *of the differential delay Eq.* (10.4) *satisfies the inequality*

$$\|\hat{Z}_1\|_{C(0,\infty)} \leq \frac{\vartheta_U}{1 - V_C(R)\vartheta_U}.$$

Proof Consider the non-homogeneous equation

$$\dot{w} - Ew = f \quad (f \in C(0, \infty)).$$

It is equivalent to the following one:

$$w(t) = \int_0^t U(t,s)(Ew(s) + f(s)) ds.$$

Hence,

$$\|w\|_{C(0,\infty)} \leq \vartheta_U (\|wx\|_{C(0,\infty)} + \|f\|_{C(0,\infty)}).$$

We thus we arrive at the inequality

$$\|w\|_{C(0,\infty)} \leq \vartheta_U (V_C(R)\|w\|_{C(0,\infty)} + \|f\|_{C(0,\infty)}).$$

Now condition (10.8) ensures the required result. □

Lemma 6.15 *Let the conditions* (10.5) *and* $q_0\hat{\chi}_0 < 1$ *hold. Then*

$$\vartheta_U < \frac{\varepsilon_A}{1 - q_0\hat{\chi}_0}.$$

Proof Consider the differential equation

$$\dot{x}(t) = A(t)x(t) + f(t) \tag{10.9}$$

with the zero initial condition $x(0) = 0$. Rewrite it as

$$\dot{x}(t) = A(s)x(t) + (A(t) - A(s))x(t) + f(t).$$

Hence

$$x(t) = \int_0^t e^{A(s)(t-t_1)}[(A(t_1) - A(s))x(t_1) + f(t_1)]dt_1.$$

Take $s = t$. Then

$$\|x(t)\|_n \le \int_0^t \|e^{A(t)(t-t_1)}\| \|(A(t_1) - A(t))x(t_1)\| dt_1 + c_0,$$

where

$$c_0 := \sup_{s,t} \int_0^t \|e^{A(s)(t-t_1)}\|_n \|f(t_1)\|_n dt_1$$

$$\le \|f\|_{C(0,\infty)} \sup_s \int_0^\infty \|e^{A(s)t_1}\|_n dt_1 \le \varepsilon_A \|f\|_{C(0,\infty)}.$$

Thus, for any $T < \infty$, by (10.5) we get

$$\sup_{t\le T} \|x(t)\|_n \le c_0 + q_0 \sup_{t\le T} \|x(t)\|_n \int_0^T \|e^{A(t)(T-t_1)}\| |t_1 - T| dt_1$$

$$\le c_0 + q_0 \sup_{t\le T} \|x(t)\|_n \int_0^T \|e^{A(t)u}\| u du \le c_0 + q_0\hat{\chi}_0 \sup_{t\le T} \|x(t)\|_n.$$

So

$$\|x\|_{C(0,T)} \le \frac{c_0}{1 - q_0\hat{\chi}_0} = \frac{\varepsilon_A \|f\|_{C(0,\infty)}}{1 - q_0\hat{\chi}_0}.$$

Hence, letting $T \to \infty$, we get

$$\|x\|_{C(0,\infty)} \leq \frac{c_0}{1 - q_0 \hat{\chi}_0} = \frac{\varepsilon_A \|f\|_{C(0,\infty)}}{1 - q_0 \hat{\chi}_0}.$$

This proves the lemma. □

Proof of Lemma 6.13 The required result at once follows from Lemma 6.14 and the previous lemma. □

Lemma 6.13 and Corollary 6.5 imply the main result of this section.

Theorem 6.6 *Let* V_1 *be defined by* (9.6). *Assume that the conditions* (1.3a), (10.5)–(10.7) *and*

$$((V_C(R) + \|A(.)\|_C)\tau(A, E) + 1)V_1 < 1$$

hold. Then the neutral type Eq. (10.1) *is exponentially stable.*

Remark 6.4 Estimates for ε_A and $\hat{\chi}_0$, under various assumptions can be taken from Sects. 2.5, 2.7 and 2.9.

6.11 Comments

The material of this chapter is based on the papers Gil' (2013b, 2014a).

Recall that the Bohl-Perron principle means that the homogeneous ordinary differential equation (ODE) $\dot{y} = A(t)y$ $(t \geq 0)$ with a variable $n \times n$-matrix $A(t)$, bounded on $[0, \infty)$ is exponentially stable, provided the nonhomogeneous ODE $\dot{x} = A(t)x + f(t)$ with the zero initial condition has a bounded solution for any bounded vector valued function f, cf. (Daleckii and Krein, 1971).

In the book by Halanay (1966, Theorem 4.15) the Bohl-Perron principle was generalized to a class of retarded systems with finite delays; besides the asymptotic (not exponential) stability was proved. The result from Halanay (1966) was afterwards considerably developed, cf. the book Azbelev and Simonov (2003) and interesting papers Berezansky and Braverman (2009), Berezansky et al. (2008), in which the generalized Bohl-Perron principle was effectively used for the stability analysis of the first and second order scalar equations. In particular, in Berezansky and Braverman (2009), the scalar non-autonomous linear functional differential equation $\dot{x}(t) + a(t)x(h(t)) = 0$ is considered. The authors give sharp conditions for exponential stability, which are suitable in the case that the coefficient function $a(t)$ is periodic, almost periodic or asymptotically almost periodic, as often encountered in applications. In the paper Berezansky et al. (2008), the authors provide sufficient conditions for the stability of rather general second-order delay differential equations.

In Gil' (2013b) the generalized Bohl-Perron principle has been extended to neutral type equations. In the paper Gil' (2011b) (see also Gil' (2013a)) a result similar to the Bohl-Perron principle has been derived in terms of the norm of the space L^p, which is called the L^p-version of the generalized Bohl-Perron principle.

Furthermore, the study of the time-variant neutral functional differential equations in many cases is based on the questions of the action of difference operators of the type $(Ef)(t) = A(t)f(\tau(t)), \tau(t) \leq t \ (t \geq 0)$ with in the spaces of discontinuous functions, for example, in the spaces of summable or essentially bounded functions. The problem of action of the pointed operators is important in the context of this chapter. In order to achieve the action of the considered operator in the space of essentially bounded or summable functions one has assume that $mes\{t : \tau(t) = c\} = 0$ for every constant c. This is necessary condition. cf. the interesting papers Drakhlin (1986) and Drakhlin and Plyshevskaya (1978). Sufficient conditions are written in these two papers via the Radon derivatives. In this connection, see also the book Azbelev et al. (1995), where the similar problems are discussed.

Note that the idea to rewrite the neutral differential equation of the type (1.1) in the form

$$\dot{x} = (I - E_1)^{-1} E_2 x$$

is not new. The similar idea was used in the paper Domoshnitsky (1983) to study boundary value problems and oscillation properties of scalar equations. Then in the paper Gusarenko and Domoshnitsk6 (1989), this idea has been used to prove the non-oscillations of a first order neutral equation. In that paper some tests of the exponential stability are also presented. The material of the paper Bainov and Domoshnitsky (1993) is close to the part of the present chapter about perturbations of autonomous systems.

Chapter 7
Nonlinear Vector NDEs

The present chapter deals with the stability of solutions to vector functional differential equations with autonomous neutral type linear parts and nonlinear causal mappings. Explicit conditions for the exponential, absolute and input-to-state stabilities are derived. Moreover, solution estimates for the considered equations are established. They provide us bounds for the regions of attraction of steady states.

Again, $C^1(-\eta, 0) = C^1([-\eta, 0], \mathbf{C}^n)$ $(0 < \eta < \infty)$, $C(a, b) = C([a, b], \mathbf{C}^n)$ and $L^p(a, b) = L^p([a, b], \mathbf{C}^n)$.

7.1 Causal Mappings

Let $X(a, \infty)$ $(-\infty < a \leq 0)$ be a Banach space of functions defined on $[a, \infty)$ with values in $\mathbf{C}^n$ and the unit operator I. For example, $X(a, \infty) = C([a, \infty), \mathbf{C}^n)$ or $X(a, \infty) = L^p([a, \infty), \mathbf{C}^n)$. For all $\tau > 0$ introduce the subspaces $X(a, \tau)$ by

$$X(a, \tau) := \{f \in X(a, \infty) : f(t) \equiv 0,\ t > \tau\}.$$

Besides, we put $\|f\|_{X(a,\tau)} = \|f\|_{X(a,\infty)}$ for all $f \in X(a, \tau)$. Denote by $P_\tau (0 < \tau < \infty)$ the projections of $X(0, \infty)$ onto $X(0, \tau)$. That is,

$$(P_\tau w)(t) = \begin{cases} w(t) & \text{if } 0 \leq t \leq \tau, \\ 0 & \text{if } t > \tau \end{cases} \qquad (w \in X(0, \infty)).$$

In addition, $P_0 = 0$, and $P_\infty w = w$.

Definition 7.1 Let $\Omega_X \subseteq X(-\eta, \infty)$ be a closed domain containing origin and F be a continuous mapping of Ω_X into $X(0, \infty)$, satisfying the condition

$$F0 = 0, \tag{1.1}$$

M.I. Gil', *Stability of Neutral Functional Differential Equations*,
Atlantis Studies in Differential Equations 3, DOI 10.2991/978-94-6239-091-1_7

and for any $u \in \Omega_X$ and all $\tau > 0$ one has

$$P_\tau F u = P_\tau F u_\tau, \tag{1.2}$$

where

$$u_\tau(t) = \begin{cases} u(t) & \text{if } -\eta \le t \le \tau, \\ 0 & \text{if } t > \tau. \end{cases}$$

Then F will be called a (continuous) causal mapping (operator).

Lemma 7.1 *Let F be a causal mapping acting from Ω_X into $X(0,\infty)$, and*

$$\|Fw\|_{X(0,\infty)} \le q\|w\|_{X(-\eta,\infty)} \quad (q = const > 0;\ w \in \Omega_X).$$

Then for all $T > 0$, one has $\|Fw\|_{X(0,T)} \le q\|w\|_{X(-\eta,T)}$ and F continuously maps $\Omega_X \cap X(-\eta, T)$ into X(0, T).

Proof Put

$$w_T(t) = \begin{cases} w(t) & \text{if } -\eta \le t \le T, \\ 0 & \text{if } t > T \end{cases}$$

and $F_T = P_T F$. According to (1.2), $F_T w = F_T w_T$. Consequently,

$$\begin{aligned} \|Fw\|_{X(0,T)} &= \|F_T w\|_{X(0,\infty)} = \|F_T w_T\|_{X(0,\infty)} \\ &\le \|F w_T\|_{X(0,\infty)} \le q\|w_T\|_{X(-\eta,\infty)} = q\|w\|_{X(-\eta,T)}. \end{aligned}$$

Since F is continuous on $X(-\eta,\infty)$, the continuity of F on $X(-\eta, T)$ is obvious. This proves the result. □

Now let $X(a,\infty) = C(a,\infty)$ and

$$\Omega(\rho) = \Omega_C(\rho) = \{w \in C(-\eta,\infty) : \|w\|_{C(-\eta,\infty)} \le \rho\}$$

for a positive $\rho \le \infty$. The following condition is often used below:

$$\|Fw\|_{C(0,\infty)} \le q\|w\|_{C(-\eta,\infty)}\ (w \in \Omega(\rho)). \tag{1.3}$$

Example 7.1 An example of a causal mapping satisfying (1.3) is given by the expression

$[Fu](t) = \psi(u(t), u(t-\eta))$ $(t \ge 0)$, where ψ is a continuous function mapping $\mathbf{C}^n \times \mathbf{C}^n$ into $\mathbf{C}^n$ and satisfying the condition

$$\|\psi(v_1, v_2)\|_n \le b_1\|v_1\|_n + b_2\|v_2\|_n \quad (v_1, v_2 \in \mathbf{C}^n;\ \|v_1\|_n, \|v_2\|_n \le \rho)$$

with positive constants b_1, b_2. Simple calculations show that in this example condition (1.3) holds with $q = b_1 + b_2$.

Example 7.2 Another example of a causal mapping satisfying (1.3) is given by the expression

$$[Fu](t) = \int_0^\eta a(s, u(t-s))ds \quad (t \geq 0),$$

where $a(.,.)$ is a continuous function mapping $[0, \eta] \times \mathbf{C}^n$ into $\mathbf{C}^n$ and satisfying the condition

$$\|a(s, v)\|_n \leq c(s)\|v\|_n \ (z \in \mathbf{C}^n; \ \|v\|_n \leq \rho; \ s \in [0, \eta])$$

with a positive continuous function $c(s)$. Simple calculations show that in this example condition (1.2) holds with

$$q = \int_0^\eta c(s)ds.$$

7.2 Existence of Solutions

$R_0(\tau)$ and $\tilde{R}_0(\tau)$ are again real $n \times n$-matrix-valued functions defined on the segment $[0, \eta]$, whose entries have bounded variations and finite numbers of jumps and $\tilde{R}_0(\tau)$ does not have a jump at zero. *Everywhere in this section and in the next one F is a causal mapping of $\Omega(\rho)$ into $C(0, \infty)$.* Consider the problem

$$\dot{x}(t) - \int_0^\eta d\tilde{R}_0(s)\dot{x}(t-s) - \int_0^\eta dR_0(s)x(t-s) = [Fx](t) + f(t) \quad (t > 0) \quad (2.1)$$

and

$$x(t) = \phi(t) \quad \text{for } -\eta \leq t \leq 0, \quad (2.2)$$

where $\phi \in C^1(-\eta, 0)$ and $f \in C(0, \infty)$ are given.

Again suppose that the inequality

$$\|Fw\|_{C(0,\infty)} \leq q\|w\|_{C(-\eta,\infty)} \quad (w \in \Omega(\rho)) \quad (2.3)$$

holds and

$$Eu(t) = \int_0^\eta dR_0(\tau)u(t-\tau), \quad \text{and} \quad \tilde{E}u(t) = \int_0^\eta d\tilde{R}_0(\tau)u(t-\tau)$$

($u \in C(-\eta, \infty)$). Thus (2.1) can be rewritten as

$$(I - \tilde{E})\dot{x} - Ex = Fx + f.$$

A (mild) solution of problem (2.1) and (2.2) *is a continuous function* $x(t)$ satisfying the equation

$$x(t) = z(t) + \int_0^t Z(t - t_1)(Fx(t_1) + f(t_1))dt_1 \quad (t \geq 0), \tag{2.4}$$

and condition (2.2). Here $Z(t)$ is the fundamental solution of the linear equation

$$\dot{z} - \tilde{E}\dot{z} - Ez = 0, \tag{2.5}$$

and $z(t)$ is a solution of the linear problem (2.5) and (2.2). Recall that

$$Q(\lambda) = I\lambda - \lambda \int_0^\eta exp(-\lambda s)d\tilde{R}_0(s) - \int_0^\eta exp(-\lambda s)dR_0(s) \quad (\lambda \in \mathbf{C})$$

is the characteristic function to (2.5). In addition, it is assumed that

$$var(\tilde{R}_0) < 1 \tag{2.6}$$

and

$$\text{all the zeros of } det \text{ are in the open left half-plane} C_-. \tag{2.7}$$

So due to Corollary 5.3 *the linear equation* (2.5) *is exponentially stable*. Recall that the Cauchy operator to Eq. (2.5) is defined by

$$\hat{Z}w(t) = \int_0^t Z(t - t_1)w(t_1)dt_1 \quad (w \in C(0, \infty)),$$

and assume that

$$\|\hat{Z}\|_{C(0,\infty)} < \frac{1}{q} \tag{2.8}$$

and

$$\|z\|_{C(-\eta,\infty)} + \|\hat{Z}f\|_{C(0,\infty)} + q\rho\|\hat{Z}\|_{C(0,\infty)} < \rho. \tag{2.9}$$

If $\rho = \infty$, then (2.9) is automatically fulfilled.

Theorem 7.1 *Let conditions* (2.6)–(2.9) *and* (2.3) *hold. Then problem* (2.1) *and* (2.2) *has at least one solution* $x(t)$. *Moreover,*

$$\|x\|_{C(-\eta,\infty)} \leq \frac{\|z\|_{C(-\eta,\infty)} + \|\hat{Z}f\|_{C(0,\infty)}}{1 - q\|\hat{Z}\|_{C(0,\infty)}}. \tag{2.10}$$

Proof Take a fixed finite $T > 0$ and put

$$\Omega(\rho, T, \phi) = \{w \in C(-\eta, T) : \|w\|_{C(0,T)} \leq \rho;\ w(t) = \phi(t);\ -\eta \leq t \leq 0\}.$$

It is not hard to check that $\Omega(\rho, T, \phi)$ is a closed convex set. Define on $\Omega(\rho, T, \phi)$ the mapping Φ by

$$\Phi w(t) = z(t) + \int_0^t Z(t-s)[Fw + f](s)ds \quad (0 \leq t \leq T; w \in \Omega(\rho, T, \phi)),$$

and

$$\Phi w(t) = \phi(t) \quad \text{for } -\eta \leq t \leq 0.$$

It is clear that $\|\hat{Z}\|_{C(0,T)} \leq \|\hat{Z}\|_{C(0,\infty)}$ and $\|z\|_{C(-\eta,T)} \geq \|\phi\|_{C(-\eta,0)}$.

Making use (2.3) and Lemma 7.1, we have

$$\|\Phi w\|_{C(-\eta,T)} \leq \|z\|_{C(-\eta,T)} + \|\hat{Z}\|_{C(0,\infty)} q\|w\|_{C(-\eta,T)} + l,$$

where $l = \|\hat{Z}f\|_{C(0,\infty)}$. According to (2.9), Φ maps $\Omega(\rho, T, \phi)$ into itself. Operator $\hat{Z}$ is compact in $C(0, T)$ since its kernel is continuous. In addition, F is continuous and bounded, and therefore Φ is continuous and compact. Hence due the Schauder principle we get the existence of solutions. Furthermore,

$$\|x\|_{C(-\eta,T)} = \|\Phi x\|_{C(-\eta,T)} \leq \|z\|_{C(-\eta,T)} + \|\hat{Z}\|_{C(0,T)} q\|x\|_{C(-\eta,T)} + l.$$

Hence, due to (2.8), we easily obtain (2.10), completing the proof. □

7.3 Lyapunov's Stability

Consider the equation

$$\dot{x}(t) - \int_0^\eta dR_0(s)\dot{x}(t-s) - \int_0^\eta d\tilde{R}_0(s)x(t-s) = [Fx](t) \quad (t > 0). \tag{3.1}$$

In the rest of the present chapter the existence and uniqueness of the considered solutions is assumed.

For example, it is not hard to check that (3.1) has a unique solution, if the Lipschitz condition

$$\|Fw - Fw_1\|_{C(0,\infty)} \le q\|w - w_1\|_{C(-\eta,\infty)} \ (w, w_1 \in \Omega(\rho))$$

is fulfilled with the constant q independent of T, provided the conditions (2.8) and

$$\|z\|_{C(-\eta,\infty)} < (1 - q\|\hat{Z}\|_{C(0,\infty)})\rho. \tag{3.2}$$

hold.

Definition 7.2 Let F be a causal mapping from $\Omega(\rho)$ $(\rho \le \infty)$ into $C(0,\infty)$. Then the zero solution to Eq. (3.1) is said to be stable (in the Lyapunov sense), if for any $\varepsilon > 0$, there exists a positive $\delta \le \rho$, such that the inequality $\|\phi\|_{C^1(-\eta,0)} \le \delta$ implies $\|x\|_{C(0,\infty)} \le \varepsilon$ for a solution $x(t)$ of problem (3.1) and (2.2).

Theorem 7.1 implies.

Theorem 7.2 *Let F be a causal mapping from $\Omega(\rho)$ into $C(0,\infty)$, and conditions* (2.3)*, and* (2.8) *hold. Then the zero solution to Eq.* (3.1) *is stable. If, in addition, condition* (3.2) *holds, then a solution $x(t)$ of problem* (3.1) *and* (3.2) *satisfies the inequality*

$$\|x\|_{C(-\eta,\infty)} \le \frac{\|z\|_{C(-\eta,\infty)}}{1 - q\|\hat{Z}\|_{C(0,\infty)}}. \tag{3.3}$$

Since the linear equation (2.5) is exponentially stable, there is a constant c_0, such that

$$\|z\|_{C(-\eta,\infty)} \le c_0\|\phi\|_{C^1(-\eta,0)}. \tag{3.4}$$

Due to (3.4), the inequality

$$c_0\|\phi\|_{C^1(-\eta,0)} \le \rho(1 - q\|\hat{Z}\|_{C(0,\infty)})$$

gives us a bound for the region of attraction.

Furthermore, *if $F : C(-\eta,\infty) \to C(0,\infty)$ is causal and the condition*

$$\lim_{\|w\|_{C(-\eta,\infty)} \to 0} \frac{\|Fw\|_{C(0,\infty)}}{\|w\|_{C(-\eta,\infty)}} = 0 \tag{3.5}$$

holds for all $w \in C(-\eta,\infty)$ uniformly in w, then Eq. (3.1) *will be called a quasilinear causal equation.* Let us prove the stability condition in the linear approximation.

Theorem 7.3 *Let* (3.1) *be a quasilinear causal equation, and conditions* (2.6) *and* (2.7) *hold. Then the zero solution to Eq.* (3.1) *is stable.*

Proof From (3.5) it follows that for any $\rho > 0$, there is a $q > 0$, such that (2.3) holds, and $q = q(\rho) \to 0$ as $\rho \to 0$. Take ρ in such a way that the condition $q\|\hat{Z}\|_{C(0,\infty)} < 1$ is fulfilled. Now the required result is due the to the previous theorem. □

For instance, if

$$\|[Fw](t)\|_n \leq \sum_{k=1}^{m} \int_0^{\eta} \|w(t-s)\|_n^{p_k} d\mu_k(s) \quad (t \geq 0;\ w \in C(-\eta, \infty)), \tag{3.6}$$

where $\mu_k(s)$ are nondecreasing functions, and $p_k = const \geq 1$, then (3.5) holds. Indeed, we have

$$\|Fw\|_{C(0,\infty)} \leq \sum_{k=1}^{m} var(\mu_k)\|w\|_{C(-\eta,\infty)}^{p_k}.$$

So for any finite $\rho > 0$ we obtain (3.5) with

$$q = q(\rho) = \sum_{k=1}^{m} \rho^{p_k - 1} var(\mu_k).$$

Recall that that estimates for $\|\hat{Z}\|_{C(0,\infty)}$ can be found, for instance, in Sects. 5.7 and 5.11.

7.4 L^2-Absolute Stability of Nonlinear NDEs

Again consider Eq. (3.1), assuming now that F is a causal mapping of $L^2(-\eta, \infty)$ into $L^2(0, \infty)$. It is assumed that there is a constant q_2, such that

$$\|Fy\|_{L^2(0,\infty)} \leq q_2\|y\|_{L^2(-\eta,\infty)} \quad (y \in L^2(-\eta, \infty)). \tag{4.1}$$

A (mild) solution is defined as in Sect. 7.2.

We will say that Eq. (3.1) is *absolutely L^2-stable in the class of the nonlinearities satisfying* (4.1), if it has a unique solution and there is a positive constant $\hat{m}$ independent of the specific form of F (but dependent on q_2), such that

$$\|x\|_{L^2(0,\infty)} \leq \hat{m}\|\phi\|_{C^1(-\eta,0)} \tag{4.2}$$

for any solution $x(t)$ of problem (3.1) and (2.2).

A simple example of a causal mapping satisfying (4.1) is given by the expression

$$[Fy](t) = \psi(y(t), y(t-\eta)) \quad (t \geq 0),$$

where ψ is a continuous function mapping of $\mathbf{C}^n \times \mathbf{C}^n$ into $\mathbf{C}^n$ and satisfying the condition

$$\|\psi(u_1, u_2)\|_n \le b_1\|u_1\|_n + b_2\|u_2\|_n \quad (u_1, u_2 \in \mathbf{C}^n)$$

with positive constants b_1, b_2. Indeed, in this case

$$\int_0^\infty \|[Fy](t)\|_n^2 dt \le \int_0^\infty (b_1\|y(t)\|_n + b_2\|y(t-\eta)\|_n)^2 dt$$
$$\le 2\int_0^\infty (b_1^2\|y(t)\|_n^2 + b_2^2\|y(t-\eta)\|_n^2) dt$$
$$\le 2\int_{-\eta}^\infty (b_1^2 + b_2^2)\|y(t)\|_n^2 dt.$$

So in this example condition (4.1) holds with

$$q_2 = \sqrt{2(b_1^2 + b_2^2)}.$$

Furthermore, recall that $\theta(Q)$ is defined in Sect. 5.3.

Lemma 7.2 *Let the conditions* (2.6) *and* (2.7) *and*

$$q_2\theta(Q) < 1 \tag{4.3}$$

hold. Then Eq. (3.1) *is absolutely L^2-stable in the class of the nonlinearities satisfying* (4.1). *Moreover, any solution $x(t)$ of problem* (3.1) *and* (2.2) *satisfies the inequality*

$$\|x\|_{L^2(-\eta,\infty)} \le \frac{\|z\|_{L^2(-\eta,\infty)}}{1 - q_2\theta(Q)}, \tag{4.4}$$

where $z(t)$ is a solution of the linear problem (2.5) *and* (2.2).

Proof Take a finite $T > 0$ and recall that due to Theorem 5.1,

$$\|\hat{Z}\|_{L^2(0,\infty)} \le \theta(Q). \tag{4.5}$$

It is clear that $\|\hat{Z}\|_{L^2(0,T)} \le \|\hat{Z}\|_{L^2(0,\infty)}$ and $\|z\|_{L^2(-\eta,T)} \ge \|\phi\|_{L^2(-\eta,0)}$. Introduce the set

$$L^2_\phi(-\eta, T) := \{u \in L^2(-\eta, T) : u(t) = \phi(t)\ (-\eta \le t \le 0)\}.$$

It is not hard to check that $L^2_\phi(-\eta, T)$ is a closed convex set.

According to (2.4), on $L^2_\phi(-\eta, T)$ define the mapping Φ by

$$\Phi w(t) = z(t) + [\hat{Z} F w](t) \quad (0 \le t \le T; w \in L^2_\phi(-\eta, T)),$$

and $\Phi w(t) = \phi(t)$ for $-\eta \le t \le 0$. Then by (4.1) and (4.5),

$$\|\Phi w\|_{L^2(-\eta,T)} \le \|z\|_{L^2(\eta,T)} + \theta(Q) q_2 \|w\|_{L^2(-\eta,T)}.$$

So for a sufficiently large $r > 0$, Φ maps the set

$$\{u \in L^2_\phi(-\eta, T) : \|u\|_{L^2(-\eta,T)} \le r\}$$

into itself. Taking into account that Φ is compact, due to the Schauder fixed point theorem, we prove the existence of solutions. Furthermore, we have

$$\|x\|_{L^2(-\eta,T)} = \|\Phi x\|_{L^2(-\eta,T)} \le \|z\|_{L^2(-\eta,T)} + \theta(Q) q_2 \|x\|_{L^2(-\eta,T)}.$$

Hence ,

$$\|x\|_{L^2(-\eta,T)} \le \frac{\|z\|_{L^2(-\eta,T)}}{1 - q_2 \theta(Q)}.$$

Now letting $T \to \infty$ we obtain the required inequality (4.4). From (2.4) with $f \equiv 0$ it follows that $x(t)$ is continuous.

Since (2.5) is stable, there is a constant m_1, such that

$$\|z\|_{L^2(0,\infty)} \le m_1 \|\phi\|_{C^1(-\eta,0)}.$$

This and (4.4) proves the absolute L^2-stability, completing the proof. □

Recall that

$$g(A) = (N_2^2(A) - \sum_{k=1}^n |\lambda_k(A)|^2)^{1/2}.$$

Besides,

$$g^2(A) \le N_2^2(A - A^*)/2 \quad \text{and} \quad g(e^{i\tau} A + zI) = g(A) \tag{4.6}$$

for all $\tau \in \mathbf{R}$ and $z \in \mathbf{C}$ (see Sect. 2.1). Remind also that

$$B(z) = z \int_0^\eta exp(-zs) d\tilde{R}_0(s) + \int_0^\eta exp(-zs) dR_0(s).$$

Set

$$\Gamma(Q(z)) = \sum_{k=0}^{n-1} \frac{g^k(B(z))}{\sqrt{k!}\rho^{k+1}(Q(z))}.$$

Here $\rho(Q(z))$ is the smallest absolute value of the eigenvalues of $Q(z)$:

$$\rho(Q(z)) = \min_{k=1,\dots,n} |\lambda_k(Q(z))|.$$

As it was shown in Sect. 5.4, the inequality

$$\theta(Q) \le \Gamma_0(Q), \quad \text{where } \Gamma_0(Q) = \sup_{-\zeta_0 \le \omega \le \zeta_0} \Gamma(Q(i\omega)) \tag{4.7}$$

is valid with

$$\zeta_0 = \frac{2var(R_0)}{1 - var(\tilde{R}_0)}.$$

Now Lemma 7.2 implies

Theorem 7.4 *Let the conditions* (2.6), (2.7) *and*

$$q_2 \Gamma_0(Q) < 1 \tag{4.8}$$

hold. Then Eq. (3.1) *is absolutely L^2-stable in the class of the nonlinearities satisfying* (4.1).

Denote

$$\hat{g}(B) := \sup_{\omega \in [-\zeta_0, \zeta_0]} g(B(i\omega)) \quad \text{and} \quad \hat{\rho}(Q) := \inf_{\omega \in [-\zeta_0, \zeta_0]} \rho(Q(i\omega)).$$

Then we have

$$\Gamma_0(Q) \le \hat{\Gamma}(Q), \quad \text{where} \hat{\Gamma}(Q) := \sum_{k=0}^{n-1} \frac{\hat{g}^k(B)}{\sqrt{k!}\hat{\rho}^{k+1}(Q)}.$$

So in the previous theorem one can replace $\Gamma_0(Q)$ for $\hat{\Gamma}_0(Q)$. About various inequalities for $\Gamma_0(Q)$ and $\hat{\Gamma}(Q)$ see Chap. 5.

7.5 Discrete Delays in Linear Parts

Let $\tilde{A} = (\tilde{a}_{jk})$, $A = (a_{jk})$ and $C = (c_{jk})$ be $n \times n$-matrices. Consider the equation

$$\dot{y}(t) - \tilde{A}\dot{y}(t - \tilde{h}) + Ay(t - h) + Cy(t) = [Fy](t) \quad (t > 0), \tag{5.1}$$

where $\tilde{h}$ and h are positive constants, F is the same as in the previous section. Assume that

$$\|\tilde{A}\|_n < 1. \tag{5.2}$$

In the considered case

$$Q(z) = z(I - \tilde{A}e^{-\tilde{h}z}) + Ae^{-hz} + C \quad (z \in \mathbf{C}).$$

The entries of Q are

$$k_{jk}(z) = z(1 - \tilde{a}_{jk}e^{-\tilde{h}z}) + a_{jk}e^{-hz} + c_{jk} \quad (j, k = 1, \dots, n).$$

Thanks to the Gerschgorin criterion (see Sect. 2.11) any eigenvalue λ of $Q(z)$ satisfies the inequality

$$|k_{jj}(z) - \lambda| \le \sum_{m=1,\, m \ne j}^{n} |k_{jm}(z)| \quad (j = 1, \dots, n).$$

Hence

$$\rho(Q(z)) \ge \min_{j=1,\dots,n} \Big(|k_{jj}(z)| - \sum_{m=1,\, m\ne j}^{n} |k_{jm}(z)|\Big), \tag{5.3}$$

provided the right-hand part is positive. Furthermore, in the case (5.1) we have

$$var(\tilde{R}_0) = \|\tilde{A}\|_n, \quad var(R_0) \le \|A\|_n + \|C\|_n$$

and

$$\zeta_0 \le \frac{2(\|A\| + \|C\|)}{1 - \|\tilde{A}\|}.$$

In addition,

$$g(Q(z)) = g(B(z)) = g(z\tilde{A}e^{-\tilde{h}z} - Ae^{-hz} - C).$$

Due to (4.6),

$$\begin{aligned} g(B(i\omega)) &\le \frac{1}{\sqrt{2}} N_2(B(i\omega) - B^*(i\omega)) \\ &\le \frac{1}{\sqrt{2}}[|\omega| N_2(e^{-i\tilde{h}\omega}\tilde{A} + e^{i\tilde{h}\omega}\tilde{A}^*) + N_2(e^{-ih\omega}A - e^{i\tilde{h}\omega}A^*) + N_2(C - C^*)]. \end{aligned}$$

One can also use the relation $g(B(i\omega)) = g(e^{is}B(i\omega))$ for all real s and ω. In particular, taking $s = \tilde{h} + \pi/2$, we have

$$g(B(i\omega)) \leq \frac{1}{\sqrt{2}}[|\omega| N_2(\tilde{A} - \tilde{A}^*) + N_2(e^{-i(h-\tilde{h})\omega} A + e^{i(h-\tilde{h})\omega} A^*) + N_2(Ce^{i\tilde{h}\omega} + e^{-i\tilde{h}\omega} C^*)].$$

If $\tilde{A}$ is self-adjoint, then

$$g(B(i\omega)) \leq \frac{1}{\sqrt{2}}[N_2(e^{-i(h-\tilde{h})\omega} A + e^{i(h-\tilde{h})\omega} A^*) + N_2(Ce^{i\tilde{h}\omega} + e^{-i\tilde{h}\omega} C^*)].$$

Consequently,

$$g(B(i\omega)) \leq \sqrt{2}[N_2(A) + N_2(C)] \quad (\omega \in \mathbf{R}). \tag{5.4}$$

Example 7.3 Consider the system

$$\dot{y}_j(t) - \tilde{a}_{jj}\dot{y}_j(t - \tilde{h}) + \sum_{k=1}^{n}(a_{jk}y_k(t-h) + c_{jk}y_k(t)) = [Fy]_j(t), \tag{5.5}$$

$(j = 1, \dots, n;\ t \geq 0)$, where $[Fy]_j(t)$ are coordinates of $[Fy](t)$, and suppose that

$$0 \leq \tilde{a}_{jj} < 1, a_{jj}, c_{jj} \geq 0 \quad (j = 1, \dots, n). \tag{5.6}$$

So $\tilde{A} = diag\,(\tilde{a}_{jj})$. Then (5.3) implies

$$\rho(Q(i\omega)) \geq \min_{j=1,\dots,n} (|k_{jj}(i\omega)| - \sum_{m=1,\, m\neq j}^{n} (|a_{jm}| + |c_{jm}|)). \tag{5.7}$$

Taking into account that $|\tilde{a}_{jj}| \leq \|\tilde{A}\|_n < 1$, put

$$v_j = \frac{2(a_{jj} + c_{jj})}{1 - \tilde{a}_{jj}}$$

and assume that

$$v_j max\{h, \tilde{h}\} < \pi/2 \quad \text{and} \quad d_j := a_{jj}cos(v_j h) + c_{jj} - v_j\tilde{a}sin(v_j\tilde{h}) > 0 \tag{5.8}$$

$(j = 1, \dots, n)$. Then by Corollary 5.4 all the zeros of $k_{jj}(.)$ are in C_- and

$$\inf_{-\infty\leq\omega\leq\infty} |k_{jj}(i\omega)| \geq d_j > 0.$$

In addition, let

$$\hat{\rho}_d := \min_{j=1,\dots,n} (d_j - \sum_{m=1,\, m\neq j}^{n} (|a_{jm}| + |c_{jm}|)) > 0, \tag{5.9}$$

then according to (5.4) we get

$$\Gamma_0(Q) \le \Gamma_d := \sum_{k=0}^{n-1} \frac{(\sqrt{2}(N_2(A) + N_2(C)))^k}{\sqrt{k!}\hat{\rho}_d^{k+1}}.$$

Now Theorem 7.4 yields our next result.

Corollary 7.1 *Let conditions* (5.6), (5.8) *and* (5.9) *be fulfilled. Then system* (5.5) *is absolutely* L^2*-stable in the class of the nonlinearities satisfying* (4.1), *provided* $q_2\Gamma_d < 1$.

7.6 Distributed Delays in Linear Parts

Let us consider the equation

$$\dot{y}(t) - \tilde{A}\int_0^\eta \dot{y}(t-s)d\tilde{\mu}(s) + A\int_0^\eta y(t-s)d\mu(s) = [Fy](t) \quad (t > 0), \qquad (6.1)$$

where $F : L^2(-\eta, \infty) \to L^2(0, \infty)$ is a causal operator, $\tilde{A}$ and A are $n \times n$-matrices, and μ, $\tilde{\mu}$ are scalar nondecreasing functions, with finite numbers of jumps, $\tilde{\mu}$ does not have a jump at zero.

So in the considered case $R_0(s) = \mu(s)A$, $\tilde{R}_0(s) = \tilde{\mu}(s)\tilde{A}$, and $Q(z) = zI - B(z)$ with

$$B(z) = z\tilde{A}\int_0^\eta e^{-zs}d\tilde{\mu}(s) - A\int_0^\eta e^{-zs}d\mu(s).$$

Without loss of generality suppose that

$$var(\mu) = var(\tilde{\mu}) = 1. \qquad (6.2)$$

It is assumed that condition (5.2) holds. We have $var(R_0) = \|A\|_n$, $var(\tilde{R}_0) = \|\tilde{A}\|_n$ and

$$\zeta_0 = \frac{2\|A\|_n}{1 - \|\tilde{A}\|_n}.$$

Moreover, the definition of $g(.)$ implies

$$g(B(i\omega)) \le N_2(B(i\omega)) \le N_2(A) + |\omega| N_2(\tilde{A}).$$

So

$$g(B(i\omega)) \le \hat{g}(Q) = \zeta_0 N_2(\tilde{A}) + N_2(A) \quad (|\omega| \le \zeta_0). \qquad (6.3)$$

In particular, if

$$Q(z) = zI - z\tilde{A}e^{-z\tilde{h}} + A\int_0^\eta e^{-zs}d\mu(s), \tag{6.4}$$

then inequality (6.3) can be improved. Namely, due to (4.6) we can write

$$\begin{aligned} g(B(i\omega)) &= g(ie^{i\omega\tilde{h}}B(i\omega)) \\ &\le \frac{1}{\sqrt{2}}\left[|\omega|N_2(\tilde{A}-\tilde{A}^*) + N_2\left(\int_0^\eta e^{-i\omega(s-\tilde{h})}d\mu(s)A\right.\right. \\ &\qquad \left.\left. + \int_0^\eta e^{i\omega(s-\tilde{h})}d\mu(s)A^*\right)\right]. \end{aligned}$$

Consequently, in the case (6.4) we get

$$\hat{g}(Q) \le \frac{\zeta_0}{\sqrt{2}}N_2(\tilde{A}-\tilde{A}^*) + \sqrt{2}N_2(A).$$

Now we can directly apply Theorem 7.4.

In the rest of this subsection we suppose that $\tilde{A}$ and A commute; so the eigenvalues of Q are

$$\lambda_j(Q(z)) = z - z\int_0^\eta e^{-zs}d\tilde{\mu}(s)\lambda_j(\tilde{A}) + \int_0^\eta e^{-zs}d\mu(s)\lambda_j(A),$$

and, due to (6.2)

$$g(B(i\omega)) \le |\omega|g(\tilde{A}) + g(A).$$

So

$$g(B(i\omega)) \le g(A,\tilde{A}) := \zeta_0 g(\tilde{A}) + g(A) \quad (\omega \in [-\zeta_0,\zeta_0]).$$

Furthermore, suppose that $\lambda_k(A)$ and $\lambda_k(\tilde{A})$ $(k = 1,\ldots,n)$ are positive and put

$$v_k = \frac{2\lambda_k(A)}{1-\lambda_k(\tilde{A})}.$$

If

$$\eta v_k < \pi/2 \tag{6.5a}$$

and

$$d_k(\mu,\tilde{\mu}) := \lambda_k(A)\int_0^\eta cos\ (\tau v_k)d\mu - v_k\lambda_k(\tilde{A})\int_0^\eta sin\ (\tau v_k)d\tilde{\mu} > 0, \tag{6.5b}$$

($k = 1, \dots, n$), then by Lemma 5.12 all the characteristic values of Q are in C_- and

$$\inf_{\omega \in \mathbf{R}} |\lambda_j(Q(i\omega))| \geq \tilde{d}_{com} := \min_k d_k(\mu, \tilde{\mu}) \ \ (j = 1, \dots, n).$$

So

$$\hat{\Gamma}(Q) \leq \Gamma_{com}(Q) := \sum_{k=0}^{n-1} \frac{\hat{g}^k(A, \tilde{A})}{\sqrt{k!}\tilde{d}_{com}^{k+1}}.$$

Now Theorem 7.4 implies

Corollary 7.2 *Let $\tilde{A}$ and A be commuting matrices with positive eigenvalues. Let the conditions* (5.2), (6.2), (6.5a), *and* (6.5b) *and $q_2 \Gamma_{com}(Q) < 1$ be fulfilled. Then Eq.* (6.1) *is absolutely L^2-stable in the class of the nonlinearities satisfying* (4.1).

Example 7.4 Consider the system

$$\dot{y}_j(t) - a\int_0^1 \dot{y}_j(t-s)ds + \sum_{k=1}^{2} c_{jk} y_k(t) = \int_0^t W_j(t,s) f_j(y_1(s-h_1), y_2(s-h_2))ds, \quad (6.6)$$

where $0 \leq h_1, h_2 = const < 1$, $0 < a = const < 1$ c_{jk} $(j, k = 1, 2)$ are real constants, $W_j(t, s)$ are real functions defined on $[0, \infty)^2$ with the property

$$\int_0^\infty \int_0^t |W_j(t,s)|^2 ds\, dt \leq b_j^2 < \infty \quad (b_j = const \geq 0)$$

and $f_j(z_1, z_2)$ are functions defined on $\mathbf{C}^2$ with the property

$$|f_j(z_1, z_2)| \leq \tilde{q}_{j1}|z_1| + \tilde{q}_{j2}|z_2| \quad (\tilde{q}_{jk} = const;\ z_j \in \mathbf{C};\ j, k = 1, 2).$$

Obviously,

$$\begin{aligned}
&\int_0^\infty \Big(\int_0^t W_j(t,s) f_j(y_1(s-h_1), y_2(s-h_2))ds\Big)^2 dt \\
&\leq \int_0^\infty \int_0^t |W_j(t,s)|^2 ds\, dt \int_0^\infty |f_j(y_1(s-h_1), y_2(s-h_2))|^2 ds \\
&\leq b_j^2 \int_0^\infty (\tilde{q}_{j1}|y_1(s-h_1)| + \tilde{q}_{j2} y_2(s-h_2)|)^2 ds \\
&\leq b_j^2 2(\tilde{q}_{j1}^2 + \tilde{q}_{j2}^2) \int_{-1}^\infty (|y_1(s)|^2 + \tilde{q}_{j2} y_2(s)|^2) ds.
\end{aligned}$$

Consequently, condition (4.1) holds with

$$q_2^2 = 2\sum_{k=1}^{2} b_j^2 \sum_{k=1}^{2} \tilde{q}_{jk}^2. \tag{6.7}$$

Furthermore, the characteristic function of system (6.6) is

$$Q(z) = z(1 - a\int_0^1 e^{-zs})dsI + C \text{ with } C = (c_{jk})$$

So $\tilde{A} = aI$ and $B(z) = a\int_0^1 e^{-zs}dsI - C$. Since $\tilde{A}$ and C commute and $g(I) = 0$, we can write

$$g(B(z)) \le g(C) \le g_C = |c_{12} - c_{21}|.$$

Moreover, the eigenvalues of Q are

$$\lambda_j(Q(z)) = z - z\,a\int_0^1 e^{-zs}ds + \lambda_j(C).$$

Suppose $\lambda_k(C)$ $(k = 1, 2)$ are real and positive and put

$$v_k(a) = \frac{2\lambda_k(C)}{1-a}.$$

According to (6.5a) and (6.5b) assume that

$$v_k(a) < \pi/2 \text{ and } d_k(a) := \lambda_k(C) - v_k(a)a\,sin\,(v_k(a)) > 0 \quad (k = 1, 2), \tag{6.8}$$

then by Lemma 5.12, the characteristic values of system (6.6) are in C_{-}, and, in addition,

$$\inf_{\omega\in\mathbf{R}} |\lambda_j(Q(i\omega))| \ge \hat{d} := \min_{k=1,2} d_k(a)\ (j = 1, 2).$$

So

$$\hat{\Gamma}(Q) \le \tilde{\Gamma} := \frac{1}{\hat{d}}(1 + \frac{g_C}{\hat{d}}).$$

Thanks to Theorem 7.4 we can assert the following result.

Corollary 7.3 *System* (6.6) *is absolutely* L^2*-stable provided the conditions* (6.8) *and* $q_2\tilde{\Gamma} < 1$ *hold, where* q_2 *is defined by* (6.7).

7.7 Exponential Stability of Nonlinear Systems

Recall that $\Omega(\rho) = \{f \in C(-\eta, \infty) : \|f\|_{C(-\eta,\infty)} \le \rho\}$. In this section F is a continuous mapping of $\Omega(\rho)$ into $C(0, \infty)$, satisfying the following condition: there is a nondecreasing function $\nu(t)$ defined on $[0, \eta]$, such that

$$\|[Fu](t)\|_n \le \int_0^\eta \|u(t-s)\|_n d\nu(s) \quad (t \ge 0; u \in \Omega(\rho)) \tag{7.1}$$

for a given number $0 < \rho \le \infty$. Again consider the equation

$$\dot{x}(t) - \int_0^\eta d\tilde{R}_0(s)\dot{x}(t-s) - \int_0^\eta dR_0(s)x(t-s) = [Fx](t) \ (t \ge 0), \tag{7.2}$$

A (mild) solution of this equation with an initial function $\phi \in C^1(-\eta, 0)$ is defined as in Sect. 7.2.

The zero solution to Eq. (7.2) *is said to be exponentially stable*, if there are positive constants $r_0 \le \rho$, $\hat{m}$ and ε, such that for any ϕ with $\|\phi\|_{C^1(-\eta,0)} \le r_0$, problem (7.2) and (2.2) has a unique solution $x(t)$ and

$$\|x(t)\|_n \le \hat{m}e^{-\varepsilon t}\|\phi\|_{C^1(-\eta,0)} \quad (t \ge 0).$$

Recall also that the uniqueness of solutions is assumed and $\theta(Q)$ is defined in Sect. 5.3.

Theorem 7.5 *Let all the characteristic values of $Q(.)$ be in C_- and the conditions* (7.1), *$var(\tilde{R}_0) < 1$ and*

$$var(\nu)\theta(Q) < 1 \tag{7.3}$$

hold. Then the zero solution to (7.2) *is exponentially stable.*

This theorem is proved in the next section.

7.8 Proof of Theorem 7.5

Lemma 7.3 *Let condition* (7.1) *hold with $\rho = \infty$. Then*

$$\|Fu\|_{L^2(0,T)} \le var(\nu)\|u\|_{L^2(-\eta,T)} \quad (u \in L^2(-\eta, T)) \tag{8.1}$$

for all $T > 0$.

Proof Let w be a scalar positive function defined on $[-\eta, T]$. Then

$$\int_0^T \Big(\int_0^\eta w(t-\tau)dv(\tau)\Big)^2 dt = \int_0^T \Big(\int_0^\eta w(t-\tau)dv(\tau) \int_0^\eta w(t-s)dv(s)\Big)dt$$
$$= \int_0^\eta \int_0^\eta \int_0^T w(t-\tau)\, w(t-s)dt\, dv(\tau)dv(s).$$

But due to the Schwarz inequality

$$\int_0^T w(t-\tau)\, w(t-s)dt \le \Big(\int_0^T w^2(t-\tau)dt \int_0^T w^2(t_1-s)dt_1\Big)^{1/2} \le \int_{-\eta}^T w^2(t)dt.$$

So

$$\int_0^T \Big(\int_0^\eta w(t-\tau)dv(\tau)\Big)^2 dt \le \int_{-\eta}^T w^2(t)dt\; var^2(v).$$

This inequality with $w(t) = \|u(t)\|_n$ implies the required result. □

Lemma 7.4 *Let conditions* (7.1) *and* (7.3) *hold with* $\rho = \infty$. *Then for any* $\phi \in C^1(-\eta, 0)$, *problem* (7.2) *and* (2.2) *has a solution* $x(t)$ *which satisfies the inequality*

$$\|x\|_{L^2(-\eta,\infty)} \le \frac{\|z\|_{L^2(-\eta,\infty)}}{1 - var(v)\|\hat{Z}\|_{L^2(0,\infty)}}, \tag{8.2}$$

where $z(t)$ *is a solution of the linear problem* (2.5) *and* (2.2).

The proof of this lemma, in view of Lemma 7.3, is similar to the proof Lemma 7.2 with the replacement of q_2 for $var(v)$. It is left to the reader.

From (7.2) and Lemma 7.3, it follows that

$$\|\dot{x}\|_{L^2(0,\infty)} \le var(\tilde{R}_0)\|\dot{x}\|_{L^2(-\eta,\infty)} + (var(R_0) + var(v))\|x\|_{L^2(-\eta,\infty)}.$$

Or

$$\|\dot{x}\|_{L^2(0,\infty)} \le var(\tilde{R}_0)(\|\dot{x}\|_{L^2(0,\infty)} + \|\dot{x}\|_{L^2(-\eta,0)}) + (var(R_0) + var(v))\|x\|_{L^2(-\eta,\infty)}.$$

Due to (7.3) this gives

$$\|\dot{x}\|_{L^2(0,\infty)} \le \frac{1}{1 - var(\tilde{R}_0)}[(var(R_0) + var(v))\|x\|_{L^2(-\eta,\infty)} + var(\tilde{R}_0)\|\dot{\phi}\|_{L^2(-\eta,0)}].$$

So we obtain

Corollary 7.4 *Under the hypothesis of the previous lemma, the derivative of a solution to Eq. (7.2) satisfies the inequality*

$$\|\dot{x}\|_{L^2(0,\infty)} \leq \frac{1}{1 - var(\tilde{R}_0)}[(var(R_0) + var(\nu))(\|x\|_{L^2(0,\infty)} + \|\phi\|_{L^2(-\eta,0)}) + var(\tilde{R}_0)\|\dot{\phi}\|_{L^2(-\eta,0)}].$$

As it was above mentioned, under condition (2.6) and (2.7) the linear equation (2.5) is exponentially stable. So

$$\|z\|_{L^2(-\eta,\infty)} \leq c_1\|\phi\|_{C^1(-\eta,0)}.$$

Here and below c_j $(j = 1, 2, \ldots)$ are constants independent on the initial functions.

Thus

$$\|x\|_{L^2(-\eta,\infty)} \leq c_2\|\phi\|_{C^1(-\eta,0)} \quad \text{and} \quad \|\dot{x}\|_{L^2(-\eta,\infty)} \leq c_3\|\phi\|_{C^1(-\eta,0)}$$

Now Lemma 4.5 implies

$$\|x\|_{C(-\eta,\infty)} \leq \sqrt{2\|x\|_{L^2(-\eta,\infty)}\|\dot{x}\|_{L^2(-\eta,\infty)}} \leq c_4\|\phi\|_{C^1(-\eta,0)}. \tag{8.3}$$

Proof of Theorem 7.5 Again assume that $\rho = \infty$. Substituting

$$x(t) = y_\varepsilon(t)e^{-\varepsilon t} \tag{8.4}$$

with an $\varepsilon > 0$ into (7.2), we obtain the equation

$$\dot{y}_\varepsilon - \varepsilon y_\varepsilon - \tilde{E}_\varepsilon \dot{y}_\varepsilon + \varepsilon E_\varepsilon y_\varepsilon = E_\varepsilon y_\varepsilon + F_\varepsilon y_\varepsilon, \tag{8.5}$$

where

$$(\tilde{E}_\varepsilon f)(t) = \int_0^\eta e^{\varepsilon\tau} d_\tau \tilde{R}_0(t,\tau) f(t-\tau), \quad (E_\varepsilon f)(t) = \int_0^\eta e^{\varepsilon\tau} d_\tau R_0(t,\tau) f(t-\tau)$$

and

$$[F_\varepsilon f](t) = e^{\varepsilon t}[F(e^{-\varepsilon t} f)](t).$$

By (7.1) with $\rho = \infty$ we have

$$\|[F_\varepsilon f](t)\|_n \leq e^{\varepsilon t}\int_0^\eta e^{-\varepsilon(t-s)}\|f(t-s)\|_n d\nu \leq e^{\varepsilon\eta}\int_0^\eta \|f(t-s)\|_n d\nu.$$

Taking ε sufficiently small and applying our above arguments to Eq. (8.5), according to (8.3), we obtain

$$\|y_\varepsilon\|_{C(0,\infty)} \le c_\varepsilon \|\phi\|_{C^1(-\eta,0)} \quad (c_\varepsilon = const).$$

Now (8.4) implies

$$\|x(t)\|_{C(0,\infty)} \le c_\varepsilon \|\phi\|_{C^1(-\eta,0)} e^{-\varepsilon t} \quad (t \ge 0). \tag{8.6}$$

Therefore, in the case $\rho = \infty$, the theorem is proved.

Now let $\rho < \infty$. By the Urysohn theorem (see Sect. 1.3), there is a continuous scalar-valued function ψ_ρ defined on $C(0,\infty)$, such that

$$\psi_\rho(f) = 1 \quad (\|f\|_{C(0,\infty)} < \rho) \quad \text{and} \quad \psi_\rho(f) = 0 \quad (\|f\|_{C(0,\infty)} \ge \rho).$$

Put $[F_\rho f](t) = \psi_\rho(f)[Ff](t)$. Clearly, F_r satisfies (7.1) for all $u \in C(-\eta,\infty)$. Consider the equation

$$\dot{x} - \tilde{E}\dot{x} = Ex + F_\rho x. \tag{8.7}$$

The solution of problem (8.7) and (2.2) denote by x_ρ. For a sufficiently small ε, according to (8.6), we have

$$\|x_\rho(t)\|_{C(0,\infty)} \le c_\varepsilon \|\phi\|_{C^1(-\eta,0)}.$$

If we take $\|\phi\|_{C^1(-\eta,0)} \le \rho/c_\varepsilon$, then $F_\rho x_r = Fx$ and Eqs. (7.2) and (8.7) coincide. This and (8.6) prove the theorem. □

7.9 Stability Conditions Via Generalized Norms

In this section, the inequalities for real vectors and vector functions are understood in the coordinate-wise sense; the inequalities for real matrices and matrix functions are understood in the element-wise sense.

$\tilde{R}_0(\tau) = (\tilde{r}_{jk}(\tau))_{j,k=1}^n$ and $R_0(\tau) = (r_{jk}(\tau))_{j,k=1}^n$ are $n \times n$-matrix-valued functions defined as in Sect. 7.2; F is a causal mapping acting from $C(-\eta,\infty)$ into $C(0,\infty)$.

Rewrite (3.1) in the form

$$\dot{x}_j(t) - \sum_{k=1}^n \int_0^\eta \dot{x}_k(t-s)d\tilde{r}_{jk}(s) = \sum_{k=1}^n \int_0^\eta x_k(t-s)dr_{jk}(s) + [Fx]_j(t) \tag{9.1}$$

with $t > 0$; $j = 1,\dots,n$. Here $x(t) = (x_k(t))_{k=1}^n$, $[Fw]_j(t)$ mean the coordinates of the vector function $Fw(t)$ with a $w \in C([-\eta,\infty), \mathbf{C}^n)$. A (mild) solution of problem (9.1) and (2.2) is defined as in Sect. 7.2.

Furthermore, let $\hat{\rho} := (\rho_1,\dots,\rho_n)$ be a vector with positive coordinates $\rho_j < \infty$. We need the following set:

$$\tilde{\Omega}\ (\hat{\rho}) := \{v(t) = (v_j(t)) \in C([-\eta, \infty), \mathbf{C}^n) : \|v_j\|_{C([-\eta,\infty),\mathbf{C})} \le \rho_j; \quad j = 1, \dots, n\}.$$

If we introduce in $C([a,b], \mathbf{C}^n)$ the generalized norm as the vector

$$M_{[a,b]}(v) := (\|v_j\|_{C([a,b],\mathbf{C})})_{j=1}^n \quad (v(t) = (v_j(t)) \in C([a,b], \mathbf{C}^n)) \tag{9.2}$$

(see Sect. 1.7), then we can write down

$$\tilde{\Omega}\ (\hat{\rho}) := \{v \in C([-\eta, \infty), \mathbf{C}^n) : M_{[-\eta,\infty)}(v) \le \hat{\rho}\}.$$

It is assumed that there are nonnegative constants ν_{jk} $(j, k = 1, \dots, n)$, such that for any

$$w(t) = (w_j(t))_{j=1}^n \in \tilde{\Omega}\ (\hat{\rho}),$$

the inequalities

$$\|[Fw]_j\|_{C([0,\infty),\mathbf{C})} \le \sum_{k=1}^n \nu_{jk}\|w_k\|_{C([-\eta,\infty),\mathbf{C})} \quad (j = 1, \dots, n) \tag{9.3}$$

hold. In other words,

$$M_{[0,\infty)}(Fw) \le \Lambda(F) M_{[-\eta,\infty)}(w) \quad (w \in \tilde{\Omega}\ (\hat{\rho})), \tag{9.4}$$

where $\Lambda(F)$ is the matrix whose entries are ν_{jk}:

$$\Lambda(F) = (\nu_{jk})_{j,k=1}^n. \tag{9.5}$$

Lemma 7.5 *Let F be a causal mapping acting from $C(-\eta, \infty)$ into $C(0, \infty)$, and satisfying condition* (9.4). *Then*

$$M_{[0,T]}(Fw) \le \Lambda(F) M_{[-\eta,T]}(w) \quad (w \in \tilde{\Omega}\ (\hat{\rho}) \cap C([-\eta, T]), \mathbf{C}^n))$$

for all $T > 0$.

The proof of this lemma is left to the reader. It is similar to the proof of Lemma 7.1.

Under conditions (2.6) and (2.7), the linear equation (2.5) is exponentially stable. So the entries $Z_{jk}(t,s)$ of the fundamental solution $Z(t)$ of Eq. (2.5) satisfy the conditions

$$\gamma_{jk} := \int_0^\infty |Z_{jk}(s)| ds < \infty. \tag{9.6}$$

Denote by $\hat{\gamma}$ the matrix with the entries γ_{jk}:

$$\hat{\gamma} = (\gamma_{jk})_{j,k=1}^n.$$

Theorem 7.6 *Let conditions* (2.6), (2.7) *and* (9.3) *hold. If, in addition, the spectral radius* $r_s(\chi)$ *of the matrix* $\chi := \hat{\gamma}\Lambda(F)$ *is less than one, then the zero solution of system* (9.1) *is stable. Moreover, if a solution* z *of the linear problem* (2.5) *and* (2.2) *satisfies the condition*

$$M_{[-\eta,\infty)}(z) + \chi\hat{\rho} \le \hat{\rho}\,, \tag{9.7}$$

then a solution $x(t)$ *of problem* (9.1) *and* (2.2) *satisfies the inequality*

$$M_{[-\eta,\infty)}(x) \le (I-\chi)^{-1}M_{[-\eta,\infty)}(z). \tag{9.8}$$

Proof Take a finite $T > 0$ and define on $\Omega_T(\hat{\rho}) = \tilde{\Omega}\ (\hat{\rho}) \cap C(-\eta, T)$ the mapping Φ by

$$\Phi w(t) = z(t) + \int_0^t Z(t-t_1)[Fw](t_1)dt_1 \quad (0 \le t \le T;\, w \in \Omega_T(\hat{\rho})),$$

and

$$\Phi w(t) = \phi(t) \text{ for } -\eta \le t \le 0.$$

The generalized norm $M_{[-\eta,T]}(.)$ on $C(-\eta, T)$ is defined as on $C(-\eta, \infty)$.

Then by (9.4),

$$M_{[-\eta,T]}(\Phi w) \le M_{[-\eta,T]}(z) + \hat{\gamma}\Lambda(F)M_{[-\eta,T]}(w).$$

According to (9.7) Φ maps $\Omega_T(\hat{\rho})$ into itself. Taking into account that Φ is compact we prove the existence of solutions. Furthermore,

$$M_{[-\eta,T]}(x) = M_{[-\eta,T]}(\Phi x) \le M_{[-\eta,T]}(z) + \chi M_{[-\eta,T]}(x).$$

So

$$M_{[-\eta,T]}(x) \le (I-\chi)^{-1}M_{[-\eta,T]}(z).$$

Hence letting $T \to \infty$, we obtain (9.8), completing the proof. □

Note that since $\chi \ge 0$, $\hat{\rho} > 0$ and $r_s(\chi) < 1$, omitting simple calculations we have the inequality $\chi\hat{\rho} \le \hat{\rho}$.

The Lipschitz condition

$$M_{[0,\infty)}(Fw - Fw_1) \le \Lambda(F)M_{[-\eta,\infty)}(w - w_1) \quad (w_1, w \in \tilde{\Omega}\ (\hat{\rho})) \tag{9.9}$$

together with the Generalized Contraction Mapping theorem also allows us to prove the existence and uniqueness of solutions.

Note that one can use the well-known inequality

$$r_s(\chi) \le \max_j \sum_{k=1}^n \hat{\chi}_{jk}, \tag{9.10}$$

where $\hat{\chi}_{jk}$ are the entries of χ (see Sect. 2.11). About other estimates for the matrix spectral radius see the book by Krasnosel'skii et al. (1989).

7.10 Systems with Diagonal Linear Parts

In this section we consider an example illustrating Theorem 7.6.

Consider the system

$$\dot{x}_j - a_j\dot{x}(t-\tilde{h}_j) + b_j x(t-h_j) = [Fx]_j(t) \quad (j = 1, \ldots, n;\ t \geq 0), \tag{10.1}$$

where $b_j, h_j, \tilde{h}_j$ are positive constants and $0 < a_j = const < 1$. Let each of the equations

$$s = se^{\tilde{h}_j s} a_j + e^{h_j s} b_j \tag{10.2}$$

have a positive root. Then due to Lemma 5.13 the fundamental solution $Z_j(t)$ of the equation

$$\dot{z}_j - a_j\dot{z}(t-\tilde{h}_j) + b_j z(t-h_j) = 0, \tag{10.3}$$

is positive and

$$\int_0^\infty Z_j(t)dt = \frac{1}{b_j}. \tag{10.4}$$

In the considered case $Z_{jj}(t) = Z_j(t)$, $Z_{jk} = 0$ for $j \neq k$, and $\gamma_{jj} = 1/b_j$. Thus under condition (9.3), we have $\chi = (\nu_{jk}/b_j)_{j,k=1}^n$. So inequality (9.10) takes the form

$$r_s(\chi) \leq \max_j \frac{1}{b_j} \sum_{k=1}^n \nu_{jk}. \tag{10.5}$$

Now Theorem 7.6 implies

Corollary 7.5 *Let each of Eq.* (10.2) *has a positive root and condition* (9.3) *hold. If, in addition,*

$$\max_j \frac{1}{b_j} \sum_{k=1}^n \nu_{jk} < 1,$$

then the zero solution of system (10.1) *is stable.*

For instance, let

$$[Fx]_j(t) = f_j(x_1(t), x_1(t-\tilde{h}), x_1(t-h), \ldots, x_n(t), x_n(t-\tilde{h}), x_n(t-h))$$

where the functions $f_j : \mathbf{R}^{3n} \to \mathbf{R}$ are continuous and

$$|f_j(z_{01}, z_{11}, z_{21}, \dots, z_{0n}, z_{1n}, z_{2n})| \le \sum_{m=0}^{2}\sum_{k=1}^{n} q_{jmk}|z_{mk}| \tag{10.6}$$

$$(q_{jmk} = const \ge 0; \;\; z_{mk} \in \mathbf{R}; m = 0, 1, 2; \;\; j, k = 1, \dots, n).$$

Hence condition (9.3) follows with

$$\nu_{jk} = \sum_{m=0}^{2} q_{jmk}.$$

7.11 Input-to-State Stability

Let $F : L^2(-\eta, \infty) \to L^2(0, \infty)$ be a causal mapping. For an input $u \in L^2(0, \infty) = L^2([0, \infty), \mathbf{C}^n)$, consider in $\mathbf{C}^n$ the problem

$$\dot{x}(t) - \int_0^\eta d\tilde{R}_0(\tau)\dot{x}(t-\tau) = \int_0^\eta dR_0(\tau)x(t-\tau) + [Fx](t) + u(t) \quad (t > 0), \tag{11.1}$$

$$x(t) = 0 \text{ for } -\eta \le t \le 0, \tag{11.2}$$

where $x(t)$ is the state. *We will say that Eq.* (11.1) *is input-to-state L^2-stable, if for any $\varepsilon > 0$, there is a $\delta > 0$, such that $\|u\|_{L^2(0,\infty)} \le \delta$ implies $\|x\|_{L^2(0,\infty)} \le \varepsilon$ for any solution of problem* (11.1) *and* (11.2).

Again it is assumed that

$$\|Fw\|_{L^2(0,\infty)} \le q_2\|w\|_{L^2(-\eta,\infty)} \quad (w \in L^2(-\eta, \infty)). \tag{11.3}$$

Recall that $\theta(Q)$ is defined in Sect. 5.3.

Theorem 7.7 *Let the conditions* (2.6), (2.7) *and* (11.3) *and $q_2\theta(Q) < 1$ hold. Then system* (11.1) *is input-to-state L^2-stable. Moreover, any solution $x(t)$ of problem* (11.1) *and* (11.2) *satisfies the inequality*

$$\|x\|_{L^2(0,\infty)} \le \frac{\theta(Q)\|u\|_{L^2(0,\infty)}}{1 - q_2\theta(Q)}. \tag{11.4}$$

Proof Take a finite $T > 0$ and recall that due to Theorem 5.1, $\|\hat{Z}\|_{L^2(0,\infty)} \le \theta(Q)$. According to (2.4), in space $L^2(0, T)$ define the mapping Φ by

$$\Phi w(t) = f(t) + [\hat{Z}(Fw + u)](t) \quad (0 \le t \le T; w \in L^2(-\eta, T)),$$

and $\Phi w(t) = 0$ for $-\eta \le t \le 0$. Then by (11.3),

$$\|\Phi w\|_{L^2(-\eta,T)} \le \|z\|_{L^2(\eta,T)} + \theta(Q)(q_2\|w\|_{L^2(-\eta,T)} + \|u\|_{L^2(0,T)}).$$

So due to the condition $q_2\theta(Q) < 1$, Φ maps the set

$$\{u \in L^2(-\eta, T) : \|u\|_{L^2(-\eta,T)} \le r\}$$

into itself, provided r is sufficiently large. Taking into account that Φ is compact, due to the Schauder fixed point theorem, we prove the existence of solutions. Furthermore, we have

$$\|x\|_{L^2(0,T)} = \|\Phi x\|_{L^2(-\eta,T)} \le \theta(Q)(q_2\|x\|_{L^2(0,T)} + \|u\|_{L^2(0,T)}).$$

Hence ,

$$\|x\|_{L^2(-\eta,T)} = \frac{\theta(Q)\|u\|_{L^2(0,T)}}{1 - q_2\theta(Q)}.$$

Now letting $T \to \infty$ we obtain the required inequality (11.4). □

7.12 Comments

The material of this chapter is based on the papers Gil' (2012b, 2013d).

As it was above pointed, the stability of various nonlinear neutral type equations was investigated by many authors. The classical results can be found in the well-known books (Hale and Lunel 1993; Kolmanovskii and Nosov 1986 and Kolmanovskii and Myshkis 1999). About the recent results see the interesting papers Agarwal and Grace (2000), Chen et al. (2008), Nam and Phat (2009), Sun and Wang (2006), Wang et al. (2011) and Wu et al. (2004) and references given therein.

In particular, in the papers Chen et al. (2008) and Wu et al. (2004), systems with discrete delays have been investigated. The global exponential stability of periodic solutions for impulsive neutral-type neural networks with delays have been explored in Wang et al. (2011). In the just pointed books and papers causal nonlinearities were not considered.

As it was already mentioned, equations with nonlinear causal mappings include differential-delay, integro-differential and some other traditional equations in a Euclidean space. The stability theory of neutral type nonlinear functional-differential equations with causal mappings is in an early stage of development. The basic method for the stability analysis of such equations is the direct Lyapunov method, cf. (Corduneanu 2002; Lakshmikantham et al. 2009). In the paper Lupulescu (2008), a deep investigation of a Cauchy problem with a causal operator in a separable Banach space is presented. Besides, sufficient conditions are given for the existence and uniqueness of solutions and some properties of set solutions are investigated. Our definition of causal mappings is somewhat different from the definition of the causal operator suggested in Corduneanu (2002), Lakshmikantham et al. (2008) and Lupulescu (2008).

Chapter 8
Absolute Stability of Scalar NDEs

The present chapter is devoted to scalar nonlinear neutral type functional differential equations with autonomous linear parts. We derive explicit absolute L^2-stability conditions in terms of the norms of the Cauchy operators to the linear parts and the Lipschitz constants of the nonlinearities. In addition, we consider the so called generalized Aizerman problem (see the comments to this chapter).

In the present chapter $C(a, b)$ is the space of continuous scalar-valued functions defined on a finite or infinite real segment $[a, b]$ and equipped with the sup-norm; $L^2(a, b)$ is the space of scalar-valued functions defined on $[a, b]$ and equipped with the norm

$$\|v\|_{L^2(a,b)} = [\int_a^b |v(t)|^2 dt]^{1/2} \ (v \in L^2(a, b)).$$

8.1 Preliminaries

Throughout this chapter F is a (continuous) causal mapping acting from $L^2(-\eta, \infty)$ into $L^2(0, \infty)$ (see Sect. 7.1).

Furthermore, consider the scalar neutral type functional differential equation

$$x^{(n)}(t) + \sum_{k=1}^{n} \int_0^{\eta} x^{(k)}(t-s) d\mu_j(s) = [Fx](t) \ \ (t > 0), \tag{1.1}$$

where μ_j are real functions having bounded variations and the finite numbers of jumps. In addition, μ_n does not have a jump at zero.

For a given scalar function ϕ, having continuous derivatives up to nth order take the initial condition

$$x(t) = \phi(t) \ (-\eta \le t \le 0). \tag{1.2}$$

M.I. Gil', *Stability of Neutral Functional Differential Equations*,
Atlantis Studies in Differential Equations 3, DOI 10.2991/978-94-6239-091-1_8

It is assumed that the condition

$$\|Fu\|_{L^2(0,\infty)} \le q\|u\|_{L^2(-\eta,\infty)}(q = const > 0;\ u \in L^2(-\eta,\infty)) \tag{1.3}$$

is fulfilled.

Example 8.1 Let

$$[Fu](t) = \int_0^t a(t-s)f_0(u(s-\eta))ds \ \ (t \ge 0),$$

where $a(.) \in L^1(0,\infty)$, $f_0 : \mathbf{R} \to \mathbf{R}$ is continuous and

$$|f_0(v)| \le q_0|v| \ (q_0 = const > 0;\ v \in \mathbf{R}).$$

Then F is causal and due to the convolution property (see Sect. 1.3) we have

$$\|Fu\|_{L^2(0,\infty)} \le \int_0^\infty |a(s)|ds \ (\int_0^\infty |f_0(u(s-\eta))|^2 ds)^{1/2}.$$

Hence,

$$\|Fu\|_{L^2(0,\infty)} \le q_0\|a(.)||_{L^1}(\int_0^\infty |u(s-\eta)|^2 ds)^{1/2} \le q_0\|a(.)||_{L^1(0,\infty)}||u||_{L^2(-\eta,\infty)}.$$

So condition (1.3) holds with

$$q = q_0\|a(.)||_{L^1(0,\infty)}.$$

If there is a nondecreasing function $\nu(t)$ defined on $[0,\eta]$, such that

$$|[Fu](t)| \le \int_0^\eta |u(t-s)|d\nu(s)(t \ge 0; u \in C(-\eta,\infty)), \tag{1.4}$$

then due to Lemma 7.3 condition (1.3) holds with

$$q = var\ (\nu).$$

Example 8.2 Let f_1 be a continuous function mapping $\mathbf{R}^m$ into $\mathbf{R}$. Then the expression

$$[Fx](t) = f_1(x(t-h_1), x(t-h_2), \ldots, x(t-h_m)) \ \ (t \ge 0),$$

where $0 \le h_1 < \cdots < h_m \le \eta$ are constant delays, is an example of a causal mapping acting from $L^2(-\eta, \infty)$ into $L^2(0, \infty)$ and satisfying (1.3), provided

$$|f_1(v_1, \dots, v_m)| \le \sum_{k=1}^{m} q_k |v_k| \ (q_k = const \ge 0; \ v_k \in \mathbf{C}).$$

Furthermore, consider the linear equation

$$\psi^{(n)} + \sum_{k=1}^{n} \int_0^{\eta} \psi^{(k)}(t-s) d\mu_k(s) = 0 \tag{1.5}$$

and put $z_k(t) = \psi^{(n-k)}(t)$ $(k = 1, \dots, n)$. Then (1.5) is reduced to the system

$$\dot{z}_1 + \int_0^{\eta} \dot{z}_1(t-s) d\mu_n(s) + \sum_{k=1}^{n-1} \int_0^{\eta} z_{n-k}(t-s) d\mu_k(s) = 0,$$

$$\dot{z}_2 = z_1,$$

$$\dots$$

$$\dot{z}_n = z_{n-1}.$$

Rewrite it in the vector form

$$\dot{z} = \tilde{A} \int_0^{\eta} \dot{z}(t-s) d\mu_n(s) + \sum_{k=1}^{n-1} A_k \int_0^{\eta} z(t-s) d\mu_k(s) + Cz(t)$$

where C is a matrix whose entries c_{jm} are $c_{j+1,j} = 1 \quad (j = 1, \dots, n-1)$, and $c_{jm} = 0$ otherwise; $\tilde{A}$ is a matrix whose entries $\tilde{a}_{jm}$ are $\tilde{a}_{11} = -1$, and $\tilde{a}_{jm} = 0$ otherwise. Similarly matrices A_k are defined. Suppose that

$$var(\mu_n) < 1. \tag{1.6}$$

Then as it is shown in Sects. 5.1 and 5.3, one can apply the Laplace transform to Eq. (1.5) and to the corresponding non-homogeneous equation

$$u^{(n)} + \sum_{k=1}^{n} \int_0^{\eta} u^{(k)}(t-s) d\mu_k(s) = f(t) \tag{1.7}$$

with a given function f satisfying the condition $|f(t)| \le ae^{bt}$ $(t \ge 0; \ a > 0, b \in \mathbf{R})$.

Let

$$Q(\lambda) = \lambda^n + \sum_{k=1}^{n} \int_0^{\eta} \lambda^k e^{-\lambda s} d\mu_k(s)$$

be the characteristic (transfer) function of (1.7). Throughout this section it is assumed that the condition (1.6) holds and

$$\text{all the zeros of } Q(.) \text{ are in } C_- = \{z \in \mathbf{C} : Re\, z < 0\}. \tag{1.8}$$

So (due to Corollary 5.3) the linear equation (1.5) is exponentially stable.

Introduce the Green function (fundamental solution) of Eq. (1.7) as in Sect. 5.2:

$$Z(t) := \frac{1}{2\pi} \int_{-\infty}^{\infty} \frac{e^{ti\omega} d\omega}{Q(i\omega)} \quad (t \geq 0).$$

Then, applying the Laplace transform to problem (1.7) and (1.2) and omitting simple calculations we obtain the Variation of Constants formula

$$u(t) = \psi(t) + \int_0^t Z(t-s) f(s) ds \quad (t \geq 0), \tag{1.9}$$

where $\psi(t)$ is a solution of the homogeneous problem (1.5) and (1.2).

According to this formula we define *a (mild) solution of problem* (1.1) and (1.2) *as a continuous function* $x(t)$ *satisfying condition* (1.2) *and the integral equation*

$$x(t) = \psi(t) + \int_0^t Z(t-s)[Fx](s) ds \quad (t \geq 0). \tag{1.10}$$

8.2 Absolute L^2 Stability

Equation (1.1) is said to be L^2-absolutely stable (L^2-a.s.) in the class of nonlinearities (1.3), if there is a positive constant c_0 independent of the specific form of F (but dependent on q), such that

$$\|x\|_{L^2(0,\infty)} \leq c_0 \sum_{k=1}^{n} \|\phi^{(k)}\|_{C(-\eta,0)}$$

for any solution $x(t)$ of problem (1.1) and (1.2).

To formulate the stability conditions introduce the Cauchy operator to Eqs. (1.7) and (1.5) by

$$\hat{Z}f(t) = \int_0^t Z(t-s)f(s)ds \quad (t \geq 0). \tag{2.1}$$

Applying to this equality the Laplace transform we get

$$\hat{\tau}(z) = \frac{\hat{f}(z)}{Q(z)},$$

where $\hat{\tau}(z)$ and $\hat{f}(z)$ are the Laplace transforms to $\hat{Z}f(t)$ and $f(t)$, respectively.

Hence it follows

$$\|\hat{\tau}(i\omega)\|^2_{L^2(-\infty,\infty)} := \int_{-\infty}^{\infty} |\hat{\tau}(i\omega)|^2 d\omega \leq \frac{1}{\inf_{-\infty\leq\omega\leq\infty}|Q(i\omega)|^2}\|\hat{f}(i\omega)\|^2_{L^2(-\infty,\infty)}.$$

Now the Parseval equality implies

$$\|\hat{Z}f\|_{L^2(0,\infty)} \leq \frac{1}{\inf_{-\infty\leq\omega\leq\infty}|Q(i\omega)|}\|f\|_{L^2(0,\infty)}. \tag{2.2}$$

Under (1.6) put

$$a_0 = \frac{2var\ (\mu_0)}{1 - var\ (\mu_n)} \text{ and } a_k = \frac{var\ (\mu_k)}{1 - var\ (\mu_n)} \quad (k = 1, \ldots, n-1),$$

and denote by ζ_1 *the unique positive root of the algebraic equation*

$$z^n = \sum_{k=1}^{n-1} a_k z^k. \tag{2.3}$$

The quantity

$$\xi(Q) := \inf_{-\zeta_1\leq\omega\leq\zeta_1} |Q(i\omega)|$$

plays an essential role in the present section.

We need the following lemma.

Lemma 8.1 *Let condition* (1.6) *hold. Then*

$$\inf_{-\infty\leq\omega\leq\infty} |Q(i\omega)| = \xi(Q).$$

Proof We have $Q(0) = \mu_0(\eta) - \mu_0(0)$. In addition,

$$|Q(i\omega)| \geq (1 - var\ (\mu_n))|\omega|^n - |\sum_{k=1}^{n-1} \int_0^{\eta} (i\omega)^k e^{-i\omega s} d\mu_k(s)|$$

$$\geq |\omega|^n (1 - var\ (\mu_n)) - \sum_{k=1}^{n-1} |\omega|^k\ var(\mu_k) \geq var\ (\mu_0),$$

provided $|\omega| > \zeta_1$. So

$$|Q(i\omega)| \geq var(\mu_0) \geq |Q(0)| \ \ (|\omega| \geq \zeta_1).$$

Thus the minimum of $|Q(i\omega)|_n$ is attained on $[-\zeta_1, \zeta_1]$. As claimed. □

The root ζ_1 can be estimated, in particular, by Lemma 2.2 and Corollary 2.2.
Furthermore, making use (2.2) and the previous lemma, we get the following result.

Lemma 8.2 *Let conditions* (1.6) *and* (1.8) *hold. Then the Cauchy operator to Eq.* (1.7) *satisfies the inequality*

$$\|\hat{Z}\|_{L^2(0,\infty)} \leq \frac{1}{\xi(Q)}.$$

Thus, we arrive at the following result.

Theorem 8.1 *Let the conditions* (1.6) and (1.8) *and*

$$q < \xi(Q) \tag{2.4}$$

hold. Then Eq. (1.1) *is absolutely L^2-stable in the class of the nonlinearities satisfying* (1.3). *Moreover, any solution $x(t)$ of problem* (1.1) and (1.2) *satisfies the inequality*

$$\|x\|_{L^2(-\eta,\infty)} \leq \frac{\xi(Q)\|\psi\|_{L^2(-\eta,\infty)}}{\xi(Q) - q},$$

where $\psi(t)$ is a solution of the linear problem (1.5) and (1.2).

The proof of this theorem is similar to the proof of Lemma 7.2 with the replacement of $\theta(Q)$ for $1/\xi(Q)$. It is left to the reader.

8.3 The Generalized Aizerman Problem

In the present chapter we consider the following problem which we call the generalized Aizerman problem.

Problem 8.3.1 To separate a class of linear parts of Eq. (1.1), such that the exponential stability of the linear equation

$$x^{(n)}(t) + \sum_{k=1}^{n} \int_{0}^{\eta} x^{(k)}(t-s)d\mu_j(s) = \tilde{q}x(t), \tag{3.1}$$

for some $\tilde{q} \in [0, q]$ provides the L^2-absolute stability of Eq. (1.1) in the class of nonlinearities (1.3).

We need the following result.

Lemma 8.3 *If the linear equation* (1.5) *is exponentially stable and its Green's function* $Z(t)$ *is non-negative, then the characteristic function* $Q(z)$ *of* (1.5) *satisfies the relations*

$$\int_{0}^{\infty} Z(t)dt = \frac{1}{Q(0)} \tag{3.2}$$

and

$$\inf_{\omega \in \mathbf{R}} |Q(i\omega)| = Q(0). \tag{3.3}$$

Proof Taking into account that

$$\frac{1}{Q(z)} = \int_{0}^{\infty} e^{-zt} Z(t)dt$$

and $Z(t)$ is positive, we obtain

$$\frac{1}{|Q(i\omega)|} \leq \int_{0}^{\infty} Z(t)dt = \frac{1}{Q(0)} \quad (\omega \in \mathbf{R}).$$

So (3.2) and (3.3) hold. □

Theorem 8.2 *Let the linear equation* (1.5) *be exponentially stable. If, in addition, Green's function* $Z(t)$ *of* (1.5) *is non-negative and*

$$Q(0) > q, \tag{3.4}$$

then Eq. (1.1) *is* L^2*-absolutely stable in the class of nonlinearities satisfying* (1.3).

Proof In view of (3.3) we have $|Q(i\omega)| > q$ for all real ω. Now the absolute stability of (1.1) in the class of nonlinearities satisfying (1.3) is due to Theorem 8.2. This proves the theorem. □

The latter theorem separates a class of equations satisfying Problem 8.3.1 *with* $\tilde{q} = q$.

Indeed, since $|Q(i\omega)| \geq Q(0) > q$ for all real ω, and (1.5) is exponentially stable, all the roots of $Q(z)$ and $Q(z) - q$ are in the open left half plane. So due to Corollary 5.3 Eq. (3.1) with $\tilde{q} = q$ is exponentially stable.

8.4 The First Order Equations

Consider the linear equation

$$\dot{y}(t) - a\dot{y}(t - \tilde{h}) + by(t - h) = 0, \tag{4.1}$$

where $a, b, h, \tilde{h}$ are constants. Besides, $a, b > 0$ and $h, \tilde{h} \geq 0$.

Put

$$k(z) = z - zae^{-\tilde{h}z} + be^{-hz}.$$

Lemma 8.4 *Let the equation*

$$s = se^{\tilde{h}s}a + e^{hs}b \tag{4.2}$$

have a positive root ψ. *Then all the zeros of* $k(z)$ *are in* C_- *and the Green function (the fundamental solution)* $Z_1(t)$ *to* (4.1) *is nonnegative. Moreover,*

$$Z_1(t) \geq e^{-\psi t} \ (t \geq 0), \tag{4.3}$$

$\inf_{\omega \in \mathbf{R}} |k(i\omega)| = b$, $\dot{Z}_1(t) \leq 0$ *and*

$$\int_0^\infty Z_1(t)dt = \frac{1}{b}. \tag{4.4}$$

Proof For the brevity put $y(t) = Z_1(t)$. Substitute $y(t) = e^{-\psi t}z(t)$ into (4.1). Then

$$\dot{z}(t) - ae^{\tilde{h}\psi}\dot{z}(t - \tilde{h}) - \psi z(t) + a\psi e^{\tilde{h}\psi}z(t - \tilde{h}) + be^{h\psi}z(t - h) = 0. \tag{4.5}$$

First assume that

$$h \geq \tilde{h} \geq 0. \tag{4.6}$$

Rewrite (4.5) as

$$\dot{z}(t) - ae^{\tilde{h}\psi}\dot{z}(t - \tilde{h}) - \psi z(t) + a\psi e^{\psi\tilde{h}}z(t - \tilde{h}) + be^{\psi h}z(t - \tilde{h}) = be^{h\psi}(z(t - \tilde{h}) - z(t - h)).$$

Due to (4.2) we get

$$a\psi e^{\psi\tilde{h}}z(t-\tilde{h}) + be^{\psi h}z(t-\tilde{h}) = \psi z(t-\tilde{h}).$$

and thus,

$$\dot{z}(t) - ae^{\tilde{h}\psi}\dot{z}(t-\tilde{h}) - \psi(z(t) - z(t-\tilde{h})) = be^{h\psi}(z(t-\tilde{h}) - z(t-h)).$$

But $z(0) = y(0) = 1$ and $z(t) = e^{\psi t}y(t) = 0$, for $t < 0$, and therefore,

$$\begin{aligned} z(t) - ae^{\tilde{h}\psi}z(t-\tilde{h}) &= 1 + \psi\int_0^t (z(t_1) - z(t_1-\tilde{h}))dt_1 \\ &\quad + be^{h\psi}\int_0^t (z(t_1-\tilde{h}) - z(t_1-h))dt_1. \\ &= 1 + \psi\int_{t-\tilde{h}}^t z(t_1)dt_1 + be^{h\psi}\int_{t-h}^{t-\tilde{h}} z(t_1)dt_1. \end{aligned}$$

Thus,

$$z(t) - ae^{\tilde{h}\psi}z(t-\tilde{h}) = 1 + \psi\int_{t-\tilde{h}}^t z(t_1)dt_1 + be^{h\psi}\int_{t-h}^{t-\tilde{h}} z(t_1)dt_1.$$

Assume that there is a $t_0 > 0$, such that $z(t_0) = 0$. Then

$$-ae^{\tilde{h}\psi}z(t_0-\tilde{h}) = 1 + \psi\int_{t_0-\tilde{h}}^{t_0} z(t_1)dt_1 + be^{h\psi}\int_{t_0-h}^{t_0-\tilde{h}} z(t_1)dt_1. \tag{4.7}$$

But this is impossible, since the right-hand part in (4.7) is greater than one and the left-hand part is non-positive. Thus $z(t) \geq 0$ and therefore $z(t) \geq 1$ for $t \geq 0$. So

$$Z_1(t) = y(t) = e^{-\psi t}z(t) \geq e^{-\psi t}.$$

Thus in the case (4.6), inequality (4.3) is proved. Now let

$$\tilde{h} \geq h \geq 0. \tag{4.8}$$

Rewrite (4.5) as

$$\dot{z}(t) - a\dot{z}(t-\tilde{h}) - \psi z(t) + a\psi e^{\psi\tilde{h}} z(t-h) + b e^{\psi h} z(t-h) = a\psi e^{\psi\tilde{h}}(z(t-h) - z(t-\tilde{h})).$$

Due to (4.2) we get

$$a\psi e^{\psi\tilde{h}} z(t-h) + b e^{\psi h} z(t-h) = \psi z(t-h).$$

and consequently,

$$\dot{z}(t) - a\dot{z}(t-\tilde{h}) - \psi(z(t) - z(t-h)) = a\psi e^{\psi\tilde{h}}(z(t-h) - z(t-\tilde{h})).$$

But $z(0) = 1$ and $z(t) = 0$ for $t < 0$, and therefore,

$$z(t) - az(t-\tilde{h}) = 1 + \psi \int_0^t (z(t_1) - z(t_1-h))dt_1 + a\psi e^{\psi\tilde{h}} \int_0^t (z(t_1-h) - z(t_1-\tilde{h}))dt_1.$$

Hence, repeating the above arguments we again obtain (4.3).
Furthermore, due to (4.2) $a < 1$. From (4.1) and (4.3) it follows that

$$\dot{y}(t) - a\dot{y}(t-\tilde{h}) = \dot{y}(t) - S\dot{y}(t) \le 0,$$

where S is the shift operator: $Sy(t) = ay(t-\tilde{h})$, $(t > \tilde{h})$, $Sy(t) = 0$, $t \le \tilde{h}$. Clearly S is positive and its spectral radius is less than one. So the operator $I - S$, where I is the unit operator in $C(0, \infty)$, has a positive inverse and thus $\dot{y}(t) \le 0$. Consequently, $y(t) \le 1$.

Furthermore, the equalities (4.4) and

$$\inf_{\omega \in \mathbf{R}} |k(i\omega)| = b$$

are due to Lemma 8.3, as claimed. □

The previous lemma and Theorem 8.2 imply

Corollary 8.1 *Let Eq.* (4.2) *have a positive root and* $b > q$. *Then the equation*

$$\dot{y}(t) - a\dot{y}(t-\tilde{h}) + by(t-h) = [Fy](t) \tag{4.9}$$

is L^2*-absolutely stable in the class of nonlinearities satisfying* (1.3).

As it was above mentioned, if there is a positive number λ, such that

$$ae^{\tilde{h}\lambda}\lambda + be^{h\lambda} \le \lambda,$$

then due to the well-known Theorem 38.1 from the book Krasnosel'skii et al. (1989), Eq. (4.2) has a positive root $\psi \le \lambda$.

Substitute $s = \omega c$ with a positive c into (4.2). Then

$$\omega e^{\tilde{h}c\omega} a + \frac{1}{c} e^{ch\omega} b = \omega. \tag{4.10}$$

If

$$e^{\tilde{h}c} a + \frac{1}{c} e^{ch} b \leq 1, \tag{4.11}$$

then due to the mentioned Theorem 38.1 (Krasnosel'skii et al. 1989) Eq. (4.10) has a positive root which not more than one. So (4.2) has a positive root $\psi \leq c$. For example, if $c = 1/h$, then condition (4.11) takes the form

$$e^{\tilde{h}/h} a + hbe \leq 1.$$

Put

$$f(c) = e^{\tilde{h}c} a + \frac{1}{c} e^{ch} b.$$

It is simple to show that

$$c_0 = -\frac{bh}{2a\tilde{h}} + \sqrt{\left(\frac{bh}{2a\tilde{h}}\right)^2 + \frac{b}{a\tilde{h}}}$$

is the zero of $f'(c)$. Thus if

$$e^{\tilde{h}c_0} a + \frac{1}{c_0} e^{c_0 h} b \leq 1,$$

then (4.2) has a positive root.

Example 8.3 Consider Eq. (4.9) with $a = 0.3, b = 0.5, \tilde{h} = 0.1, h = 0.2$. So Eq. (4.2) has the form

$$s = 0.3se^{0.1s} + 0.5e^{0.2s}. \tag{4.12}$$

Since

$$0.3e^{0.1} + 0.5e^{0.2} < 1,$$

Equation (4.12) has a positive root. Hence, under consideration, by Corollary 8.1 Eq. (4.9) is L^2-a.s. in the class of nonlinearities (1.3), provided $q < 0.5$.

8.5 Higher Order Equations

Now consider the higher order equation

$$\prod_{j=1}^{n}(\frac{d}{dt}-a_j\tilde{E}_j\frac{d}{dt}+b_jE_j)x=Fx\ (t>0), \tag{5.1}$$

where $\tilde{E}_jx(t)=x(t-\tilde{h}_j)$, $E_jx(t)=x(t-h_j)$ ($b_j=const>0;\ h_j,\tilde{h}_j,a_j=const\geq 0;\ j=1,\dots,n$). The characteristic function of the linear part

$$\prod_{j=1}^{n}(\frac{d}{dt}-a_j\tilde{E}_j\frac{d}{dt}+b_jE_j)x=0 \tag{5.2}$$

of (5.1) is

$$Q(\lambda)=\prod_{j=1}^{n}(\lambda-\lambda a_je^{-\tilde{h}_j\lambda}+b_je^{-h_j\lambda}). \tag{5.3}$$

Let each of the equations

$$s=se^{\tilde{h}_js}a_j+e^{hs}b_j\ \ (j=1,2,\dots,n) \tag{5.4}$$

have at least one positive root. Then necessarily $a_j<1$ ($j=1,2,\dots,n$). Taking into account that a product of the Laplace transforms of several functions corresponds to the convolution of these functions, we have due to Lemma 8.4 the following result:

Lemma 8.5 *Let each of Eq.* (5.4) *have a positive root. Then the Green function of* (5.2) *is nonnegative and*

$$\inf_{\omega}|Q(i\omega)|=Q(0)=b_1b_2\dots b_n.$$

Now applying Theorem 8.2 we get

Corollary 8.2 *Let each of Eq.* (5.4) *have a positive root and* $b_1b_2\dots b_n>q$. *Then Eq.* (5.1) *is* L^2*-absolutely stable in the class of nonlinearities satisfying* (1.3).

8.6 Additional Stability Conditions

In this section we investigate scalar equations without the condition that the linear parts have positive Green's functions. To this end consider the function

$$k(z) = z\left(1 - \int_0^\eta e^{-\tau z} d\tilde{\mu}\right) + \int_0^\eta e^{-\tau z} d\mu \quad (z \in \mathbf{C}), \tag{6.1}$$

where $\mu = \mu(\tau)$ and $\tilde{\mu} = \tilde{\mu}(\tau)$ are nondecreasing functions defined on $[0, \eta]$, with

$$0 < var(\tilde{\mu}) < 1 \text{ and } var(\mu) < \infty. \tag{6.2}$$

In addition, μ and $\tilde{\mu}$ have finite numbers of jumps and $\tilde{\mu}$ does not have a jump at zero.

Put

$$v_1 = \frac{2\, var(\mu)}{1 - var(\tilde{\mu})}.$$

Lemma 8.6 *The equality*

$$\inf_{-\infty \le \omega \le \infty} |k(i\omega)| = \inf_{-v_1 \le \omega \le v_1} |k(i\omega)| \tag{6.3}$$

is valid.

Proof We have $-k(0) = var(\mu)$. In addition,

$$|k(i\omega)| \ge |\omega|(1 - var\ (\tilde{\mu})) - var\ (\mu) \ge var(\mu) \quad (\omega \in \mathbf{R}, |\omega| \ge v_1).$$

Thus the minimum of $|k(i\omega)|$ is attained on $[-v_1, v_1]$. As claimed. □

Lemma 8.7 *Let the conditions* (6.2),

$$\eta v_1 < \pi/2 \tag{6.4}$$

and

$$d_0 := \int_0^\eta cos(v_1\tau)d\mu - v_1 \int_0^\eta sin(v_1\tau)d\tilde{\mu} > 0 \tag{6.5}$$

hold. Then all the zeros of $k(.)$ *are in* C_- *and*

$$\inf_{-\infty \le \omega \le \infty} |k(i\omega)| \ge d_0 > 0. \tag{6.6}$$

Proof Clearly, $Re\ k(i\omega) = -\omega \int_0^\eta sin(\omega\tau)d\tilde{\mu} + \int_0^\eta cos(\omega\tau)d\mu$. Hence, with $|\omega| \le v_1$, we have

$$|Re\ k(i\omega)| = Re\ k(i\omega) \geq -v_1 \int_0^{\eta} sin(v_1\tau)d\tilde{\mu} + \int_0^{\eta} cos(v_1\tau)d\mu = d_0 > 0.$$

This and Lemma 8.6 prove (6.6).
Furthermore, put

$$g_2(z) = -z \int_0^{\eta} e^{-\tau z} d\tilde{\mu}$$

and $k(m, z) = \xi(z) + mg_2(z), 0 \leq m \leq 1$. where

$$\xi(z) = z + \int_0^{\eta} exp(-zs)d\mu(s)\ (z \in \mathbf{C}).$$

Due to Lemma 4.9, if $\eta\ var(\mu) < \pi/4$, then all the zeros of $\xi(z)$ are in C_- and

$$\inf_{\omega \in \mathbf{R}} |\xi(i\omega)| \geq \hat{d},$$

where

$$\hat{d} := \int_0^{\eta} cos(2var(\mu)\tau)d\mu(\tau).$$

Clearly, $v_1 > var(\mu)$ and

$$v_m := \frac{2var(\mu)}{1 - mvar(\tilde{\mu})} < v_1.$$

According to Lemma 8.6,

$$\inf_{\omega \in \mathbf{R}} |k(m, i\omega)| = \inf_{-v_1 \leq \omega \leq v_1} |k(m, i\omega)|.$$

Hence, due to (6.6),

$$|k(m, i\omega)| \geq \int_0^{\eta} cos(v_1\tau)d\mu - mv_1 \int_0^{\eta} sin(v_1\tau)d\tilde{\mu} > d_0\ (0 < m \leq 1;\ \omega \in \mathbf{R}). \tag{6.7}$$

Furthermore, assume that $k(z)$ has a zero in the closed right hand-plane $\overline{C}_+$. By the just mentioned Lemma 4.9 $k(0, z) = \xi(z)$ does not zeros on $\overline{C}_+$. So $k(m_0, i\omega)$ ($\omega \in \mathbf{R}$) should have a zero for some $m_0 \in (0, 1]$, according to continuous dependence of zeros on coefficients. But due to (6.7) this is impossible. The proof is complete. □

For instance, consider the function

$$k_1(z) = z(1 - \tilde{a}e^{-\tilde{h}z}) + ae^{-hz} + b$$

with $a, b, h, \tilde{h} = const \geq 0$, and $0 < \tilde{a} < 1$. Then

$$v_1 = \frac{2(a+b)}{1-\tilde{a}}. \tag{6.8}$$

Corollary 8.3 *Assume that v_1 is defined by* (6.8), *and the conditions*

$$hv_1 < \pi/2, \tilde{h}v_1 < \pi/2 \tag{6.9}$$

and

$$d_1 := a\ cos(v_1 h) + b - v_1\tilde{a}\ sin(v_1\tilde{h}) > 0 \tag{6.10}$$

are fulfilled. Then all the zeros of $k_1(.)$ are in C_- and

$$\inf_{-\infty\leq\omega\leq\infty} |k(i\omega)| \geq d_1 > 0.$$

Applying Theorem 8.1 and Lemma 8.7 to the equation

$$\dot{y}(t) - \int_0^\eta \dot{y}(t-\tau)d\tilde{\mu}(\tau) + \int_0^\eta y(t-\tau)d\mu(\tau) = [Fy](t), \tag{6.11}$$

we arrive at the following result.

Corollary 8.4 *Let conditions* (6.4) *and* (6.5) *hold. In addition, let*

$$q < \int_0^\eta cos(v_1\tau)d\mu - v_1 \int_0^\eta sin(v_1\tau)d\tilde{\mu}. \tag{6.12}$$

Then Eq. (6.11) *is L^2-absolutely stable in the class of nonlinearities satisfying* (1.3).

In particular, consider the equation

$$\dot{y}(t) - \tilde{a}\dot{y}(t-\tilde{h}) + ay(t-h) + by(t) = [Fy](t), \tag{6.13}$$

provided conditions (6.9) and (6.10) hold with v_1 defined by (6.8). Then Theorem 8.1 and Corollary 8.4 imply that Eq. (6.13) is L^2-absolutely stable in the class of nonlinearities satisfying (1.3), if the inequality

$$q < a\, cos(v_1 h) + b - v_1 \tilde{a}\ sin(v_1 \tilde{h})$$

is fulfilled.

8.7 Comments

The material of this chapter is based on the paper Gil' (2013c).

Let A, b, c be an $n \times n$-matrix, a column-matrix and a row-matrix, respectively. In 1949 Aizerman conjectured the following hypothesis: for the absolute stability of the zero solution of the equation $\dot{x} = Ax + bf(cx)$ in the class of nonlinearities $f : \mathbf{R} \to \mathbf{R}$, satisfying $0 \leq f(s)/s \leq q$ ($q = const > 0, s \in \mathbf{R},\ s \neq 0$) it is necessary and sufficient that the linear equation $\dot{x} = Ax + q_1 bcx$ be asymptotically stable for any $q_1 \in [0, q]$ (Aizerman 1949). This hypothesis caused the great interest among the specialists. Counterexamples were set up that demonstrated it was not, in general, true, (see Vidyasagar (1993) and references therein). In connection with these results, Voronov (1977) investigated the following problem: to find the class of systems that satisfy Aizerman's hypothesis. He also received the first important results in that direction. The author has showed that any system satisfies the Aizerman hypothesis, if its impulse function is non-negative (Gil' 1983). The similar result was proved for multivariable systems and distributed ones (Gil' 1998). On the other hand, Myshkis (1977, Section 10) pointed out at the importance of consideration of the generalized Aizerman problem for retarded systems. That problem, called the Aizerman-Myshkis problem, was investigated under various assumptions, by the author in the papers Gil' (2000b, 2002 and 2009b). The interesting results on the Aizerman-Myshkis problem can be found in the papers Popescu et al. (2004) and Rasvan (2002). In addition, the paper MacCamy and Wong (1972) should be mentioned. In that paper, absolute stability conditions for a wide class of functional and integro-differential equations have been established. These conditions are connected with the generalized Aizerman problem. For the classical results on the absolute stability of systems we refer the reader to the book Rasvan (1983) and survey Liberzon (2006).

Furthermore, as it was pointed in the note Rasvan (2013), an important aspect concerned with the paper Gil' (2013c) and also with the entire framework of the Aizerman's problem is the methodology of solving it. As it is pointed out in the book Voronov (1979), the pioneering papers of B.V. Bulgakov are important when seeking the necessary and sufficient conditions for absolute stability while the pioneering paper of Lurie and Postnikov was the starting point of the most successful approach for tackling the problem: a special Lyapunov function with the structure "quadratic form plus integral of the nonlinear function". Existence of such a function with its derivative along system's solutions being at least non-positive is reduced to the solution of the Lurie type linear matrix inequalities—equivalent to some matrix algebraic Riccati inequality. Later Popov introduced his method of the frequency domain inequalities which turned to be equivalent to the Lyapunov method via the positiveness theory relying on the Yakubovich-Kalman—Popov lemma. Due to Yakubovich and his co-workers, this equivalence is valid on Hilbert spaces also; however, even in the simplest case of time delay systems, the linear matrix inequalities become linear operator inequalities which have to be solved using adequate numerical methods. For more historic details see the above mentioned paper Rasvan (2013), in which it is also written the following: "... it is felt that the research on the Aizerman problem is an evergreen domain, with a determinant role in the analysis of the so called almost linear behavior of the dynamical systems".

Chapter 9
Bounds for Characteristic Values of NDEs

In this chapter we investigate certain properties of the characteristic values of autonomous vector NDEs. In particular, bounds for characteristic values and perturbations results are derived.

For the brevity, in this chapter $\|A\|$ denotes the spectral norm of a matrix A: $\|A\| = \|A\|_n$.

9.1 Sums of Moduli of Characteristic Values

Recall that

$$Q(z) = zI - z\int_0^\eta exp(-zs)d\tilde{R}_0(s) - \int_0^\eta exp(-zs)dR_0(s)$$

is the characteristic matrix function of the autonomous equation

$$\dot{y} - \int_0^\eta d\tilde{R}_0(s)\dot{y}(t-s) = \int_0^\eta dR_0(s)y(t-s),$$

where I is the unit matrix $R_0(s)$ and $\tilde{R}_0(s)$ are $n \times n$-matrix-valued function defined on a finite segment $[0, \eta]$, whose entries have bounded variations and finite numbers of jumps.

Enumerate the characteristic values $z_k(Q)$ of Q with their multiplicities in the nondecreasing order of their absolute values: $|z_k(Q)| \le |z_{k+1}(Q)|$ $(k = 1, 2, \ldots)$.

M.I. Gil', *Stability of Neutral Functional Differential Equations*,
Atlantis Studies in Differential Equations 3, DOI 10.2991/978-94-6239-091-1_9

The aim of this section is to estimate the sums

$$\sum_{k=1}^{j} \frac{1}{|z_k(Q)|} \quad (j = 1, 2, \ldots).$$

To this end, note that

$$Q(z) = zI - \sum_{k=1}^{\infty} \frac{(-1)^k z^{k+1}}{k!} \int_0^{\eta} s^k d\tilde{R}_0(s)$$
$$- \sum_{k=1}^{\infty} \frac{(-1)^k z^k}{k!} \int_0^{\eta} s^k dR_0(s)$$

and therefore,

$$Q(z) = \sum_{k=1}^{\infty} \frac{z^k}{k!} B_k, \tag{1.1}$$

where

$$B_0 = -\int_0^{\eta} dR_0(s) = R_0(0) - R_0(\eta),$$

$$B_1 = I - \int_0^{\eta} d\tilde{R}_0(s) + \int_0^{\eta} s dR_0(s)$$

and

$$B_k = (-1)^k \left(k \int_0^{\eta} s^{k-1} d\tilde{R}_0(s) - \int_0^{\eta} s^k dR_0(s) \right) \quad (k \geq 2).$$

In addition, it is supposed that

$$B_0 \text{ is invertible.} \tag{1.2}$$

Put $C_k = B_0^{-1} B_k$ and

$$Q_1(z) := B_0^{-1} Q(z) = \sum_{k=1}^{\infty} \frac{z^k}{k!} C_k \quad (C_0 = I).$$

Since $det\ Q_1(z) = det\ B_0^{-1}\ det\ Q(z)$, all the characteristic values of Q and Q_1 coincide.

Without loss of generality assume that

$$\eta < 1. \tag{1.3}$$

If this condition is not valid, then by the substitution $z = wa$ with some $a < 1/\eta$ into (1.1), we obtain

$$Q(aw) = awI - aw \int_0^{\eta} exp(-saw)d\tilde{R}_0(s) - \int_0^{\eta} exp(-saw)dR_0(s) = aQ_a(w),$$

where

$$Q_a(w) = wI - w \int_0^{\eta_1} exp(-\tau w)d\tilde{R}_a(\tau) - \int_0^{\eta_1} exp(-\tau w)dR_a(\tau)$$

with $\tilde{R}_a(\tau) = \tilde{R}_0(\tau/a)$, $R_a(\tau) = \frac{1}{a}R_0(\tau/a)$ and $\eta_1 = a\eta < 1$. Thus, condition (1.3) holds with $\eta = \eta_1$.

Let $r_{jk}(s)$ be the entries of $R_0(s)$. Recall that $var(R_0)$ denote the spectral norm of the matrix $\left(var(r_{jk})\right)_{j,k=1}^n$: $var(R_0) = \| \left(var(r_{jk})\right)_{j,k=1}^n \|$.

Lemma 9.1 *The inequalities*

$$\|B_k\| = \| \int_0^{\eta} s^k dR_0(s)\| \le \eta^k var\ (R_0) + k\eta^{k-1} var\ (\tilde{R}_0) \quad (k = 2, 3, \ldots)$$

are true.

Proof We have

$$\|B_k\| = \|k \int_0^{\eta} s^{k-1} d\tilde{R}_0(s) - \int_0^{\eta} s^k dR_0(s)\|$$
$$\le k \int_0^{\eta} s^{k-1} \|d\tilde{R}_0(s)\| + \int_0^{\eta} s^k \|dR_0(s)\|$$
$$\le k\eta^{k-1} \int_0^{\eta} \|d\tilde{R}_0(s)\| + \eta^k \int_0^{\eta} \|dR_0(s)\|,$$

as claimed. □

From this lemma it follows that

$$\|C_k\| \leq \|B_0^{-1}\|(\eta^k var\ (R_0) + k\eta^{k-1} var\ (\tilde{R}_0)) \quad (k \geq 2).$$

Thus under condition (1.3) we have

$$\sum_{k=1}^{\infty} \|C_k\|^2 < \infty. \tag{1.4}$$

Furthermore, denote

$$\Psi_Q := \left[\sum_{k=1}^{\infty} C_k C_k^*\right]^{1/2}.$$

So Ψ_Q is an $n \times n$-matrix. Set

$$\omega_k(Q) = \begin{cases} \lambda_k(\Psi_Q) & \text{for } k = 1, \ldots, n, \\ 0 & \text{if } k \geq n+1 \end{cases},$$

where $\lambda_k(\Psi_Q)$ are the eigenvalues of matrix Ψ_Q with their multiplicities and enumerated in the decreasing way: $\omega_k(Q) \geq \omega_{k+1}(Q)$ $(k = 1, 2, \ldots)$.

Theorem 9.1 *Let conditions* (1.2) *and* (1.3) *hold. Then the characteristic values of* Q *satisfy the inequalities*

$$\sum_{k=1}^{j} \frac{1}{|z_k(Q)|} < \sum_{k=1}^{j} \left[\omega_k(Q) + \frac{n}{k+n}\right] \quad (j = 1, 2, \ldots).$$

This result is a particular case of Theorem 12.2.1 from the book Gil' (2010a).

From the latter theorem it follows that

$$\frac{j}{|z_j(Q)|} < \sum_{k=1}^{j} \left[\omega_k(Q) + \frac{n}{k+n}\right]$$

and thus,

$$|z_j(Q)| > \frac{j}{\sum_{k=1}^{j}\left[\omega_k(Q) + \frac{n}{(k+n)}\right]} \quad (j = 1, 2, \ldots).$$

Therefore, in the disc

$$|z| \leq \frac{j}{\sum_{k=1}^{j}\left[\omega_k(Q) + \frac{n}{(k+n)}\right]} \quad (j = 2, 3, \ldots)$$

Q has no more than $j-1$ characteristic values. Let $\nu_Q(r)$ be the function counting the characteristic values of Q in the disc $|z| \leq r$. We consequently, get.

Corollary 9.1 *Let conditions* (1.2) *and* (1.3) *hold. Then the inequality* $\nu_Q(r) \leq j-1$ *is valid, provided*

$$r \leq \frac{j}{\sum_{k=1}^{j}\left[\omega_k(Q) + \frac{n}{(k+n)}\right]} \quad (j = 2, 3, \ldots).$$

Moreover, $Q(z)$ *does not have characteristic values in the disc*

$$|z| \leq \frac{1}{\lambda_1(\Psi_Q) + \frac{n}{1+n}}.$$

Let us apply the Borel (Laplace) transform. Namely, put

$$F_Q(w) = \int_0^\infty e^{-wt} Q_1(t)dt \quad (w \in \mathbf{C};\ Re\ w > 0).$$

Then

$$F_Q(w) = B_0^{-1} \int_0^\infty e^{-wt} \left(tI - t\int_0^\eta exp(-ts)d\tilde{R}_0(s) - \int_0^\eta exp(-ts)dR_0(s) \right) dt.$$

Obviously,

$$\int_0^\infty e^{-wt} t dt = \frac{1}{w^2}, \int_0^\infty e^{-wt} \int_0^\eta exp(-ts)dR_0(s)dt = \int_0^\eta \frac{1}{s+w} dR_0(s)$$

and

$$\int_0^\infty e^{-wt} t \int_0^\eta exp(-ts)d\tilde{R}_0(s)dt = \int_0^\eta \frac{1}{(s+w)^2} d\tilde{R}_0(s).$$

Consequently, we can write

$$F_Q(w) = B_0^{-1} \left[\frac{1}{w^2} I - \int_0^\eta \frac{1}{(s+w)^2} d\tilde{R}_0(s) - \int_0^\eta \frac{1}{s+w} dR_0(s) \right]. \quad (1.5)$$

In view of (1.3) one can extend this relation to the domain $|w| \geq 1$.

On the other hand

$$F_Q(w) = \sum_{k=1}^{\infty} \frac{1}{w^{k+1}} C_k$$

and therefore

$$\frac{1}{2\pi}\int_0^{2\pi} F_Q(e^{-is})F_Q^*(e^{is})ds = \sum_{k=1}^{\infty} C_k C_k^*.$$

Hence we arrive at the following result.

Lemma 9.2 *Let condition* (1.3) *hold. Then*

$$\Psi_Q^2 = \frac{1}{2\pi}\int_0^{2\pi} F_Q(e^{-is})F_Q^*(e^{is})ds$$

and consequently

$$\|\Psi_Q\|^2 \le \frac{1}{2\pi}\int_0^{2\pi} \|F_Q(e^{is})\|^2 ds \le \sup_{|w|=1} \|F_Q(w)\|^2.$$

Simple calculations, show that

$$\|\int_0^{\eta} \frac{1}{s+w} dR_0(s)\| \le \frac{var(R_0)}{1-\eta} \quad (|w|=1)$$

and

$$\|\int_0^{\eta} \frac{1}{(s+w)^2} d\tilde{R}_0(s)\| \le \frac{var(\tilde{R}_0)}{(1-\eta)^2} \quad (|w|=1).$$

Taking into account (1.5), we deduce that $\|\Psi_Q\| \le \alpha(F_Q)$, where

$$\alpha(F_Q) := \|B_0^{-1}\| \left(1 + \frac{var(\tilde{R}_0)}{(1-\eta)^2} + \frac{var(R_0)}{1-\eta}\right)$$

and consequently,

$$\omega_j(Q) \le \sup_{|w|=1} \|F_Q(w)\| \le \alpha(F_Q). \tag{1.6}$$

Note that the norm of B_0^{-1} can be estimated by the results presented in Sect. 2.3 above. Inequality (1.6) and Theorem 9.1 imply our next result.

Corollary 9.2 *Let conditions* (1.2) *and* (1.3) *hold. Then the characteristic values of* Q *satisfy the inequalities*

$$\sum_{k=1}^{j} \frac{1}{|z_k(Q)|} < j\alpha(F_Q) + n\sum_{k=1}^{j} \frac{1}{k+n} \quad (j \le n)$$

and

$$\sum_{k=1}^{j} \frac{1}{|z_k(Q)|} < n\left(\alpha(F_Q) + \sum_{k=1}^{j} \frac{1}{k+n}\right) \quad (j > n).$$

9.2 Identities for Characteristic Values

This section deals with the following sums containing the characteristic values $z_k(Q)$ $(k = 1, 2, \dots)$:

$$\tilde{s}_m(Q) := \sum_{k=1}^{\infty} \frac{1}{z_k^m(Q)} \quad (m = 2, 3, \dots),$$

$$\sum_{k=1}^{\infty} \left(Im\, \frac{1}{z_k(Q)}\right)^2 \quad \text{and} \quad \sum_{k=1}^{\infty} \left(Re\, \frac{1}{z_k(Q)}\right)^2.$$

To formulate our next result, for an integer $m \ge 2$, introduce the $m \times m$-block matrix

$$\hat{B}_m = \begin{pmatrix} -C_1 & -C_2/2 & \cdots & -C_{m-1}/(m-1)! & -C_m/m! \\ I & 0 & \cdots & 0 & 0 \\ 0 & I & \cdots & 0 & 0 \\ \cdot & \cdot & \cdots & \cdot & \cdot \\ 0 & 0 & \cdots & I & 0 \end{pmatrix} \tag{2.1}$$

whose entries are $n \times n$ matrices.

Theorem 9.2 *For any integer* $m \ge 2$, *we have*

$$\tilde{s}_m(Q) = Trace\ \hat{B}_m^m.$$

This result is a particular case of Theorem 12.7.1 from Gil' (2010a).

Furthermore, denote

$$\tau(Q) := \sum_{k=1}^{\infty} N_2^2(C_k) + (\zeta(2) - 1)n$$

and

$$\psi(Q,t) := \tau(Q) + Re\ Trace\ [e^{2it}\ (C_1^2 - C_2/2)] \quad (t \in [0, 2\pi)),$$

where

$$\zeta(z) := \sum_{k=1}^{\infty} \frac{1}{k^z} \quad (Re\ z > 1)$$

is the Riemann zeta function.

Theorem 9.3 *Let conditions* (1.2) *and* (1.3) *hold. Then for any* $t \in [0, 2\pi)$ *the relations*

$$\tau(Q) - \sum_{k=1}^{\infty} \frac{1}{|z_k(Q)|^2} = \psi(Q,t) - 2\sum_{k=1}^{\infty} \left(Re\ \frac{e^{it}}{z_k(Q)}\right)^2 \geq 0$$

are valid.

This theorem is a particular case of Theorem 12.4.1 from the book Gil' (2010a).

Note that

$$\psi(Q, \pi/2) = \tau(Q) - Re\ Trace\ (C_1^2 - \frac{1}{2}C_2)$$

and

$$\psi(Q, 0) = \tau(Q) + Re\ Trace\ (C_1^2 - \frac{1}{2}C_2).$$

Now Theorem 9.2 yields the following result.

Corollary 9.3 *Let conditions* (1.2) *and* (1.3) *hold. Then*

$$\tau(Q) - \sum_{k=1}^{\infty} \frac{1}{|z_k(Q)|^2} = \psi(Q, \pi/2) - 2\sum_{k=1}^{\infty} \left(Im\ \frac{1}{z_k(Q)}\right)^2$$
$$= \psi(Q, 0) - 2\sum_{k=1}^{\infty} \left(Re\ \frac{1}{z_k(Q)}\right)^2 \geq 0.$$

Consequently,

$$\sum_{k=1}^{\infty} \frac{1}{|z_k(Q)|^2} \leq \tau(Q),$$

$$2\sum_{k=1}^{\infty} \left(Im\ \frac{1}{z_k(Q)}\right)^2 \leq \psi(f, \pi/2)$$

and

$$2\sum_{k=1}^{\infty}\left(Re\,\frac{1}{z_k(Q)}\right)^2 \le \psi(f,0).$$

9.3 Multiplicative Representations of Characteristic Matrices

Let A be an $n\times n$-matrix and E_k $(k=1,\ldots,n)$ be the maximal chain of the invariant orthogonal projections of A. That is,

$$AE_k = E_kAE_k \ (k=1,\ldots,n) \tag{3.1}$$

and

$$0=E_0\subset E_1\subset\cdots\subset E_n = I.$$

Here and the symbol $E_{k-1}\subset E_k$ means $E_{k-1}\mathbf{C}^n\subset E_k\mathbf{C}^n$.

Besides, $\Delta E_k = E_k - E_{k-1}$ $(k=1,\ldots,n)$ are one dimensional. Set

$$\prod_{1\le k\le m}^{\rightarrow} X_k := X_1X_2\ldots X_m$$

for matrices $X_1, X_2, \ldots, X_m$. As it is shown in the book Gil' (2003, Theorem 2.9.1),

$$(I-A)^{-1} = \prod_{1\le k\le n}^{\rightarrow}\left(I+\frac{A\Delta E_k}{1-\lambda_k(A)}\right), \tag{3.2}$$

provided $I-A$ is invertible.

Furthermore, for each fixed $z\in\mathbf{C}$, $Q(z)$ possesses the maximal chain of invariant orthogonal projections, which we denote by $E_k(Q,z)$:

$$0=E_0(Q,z)\subset E_1(Q,z)\subset\cdots\subset E_n(Q,z)=I$$

and

$$Q(z)E_k(Q,z)=E_k(Q,z)Q(z)E_k(Q,z) \ (k=1,\ldots,n).$$

Moreover,

$$\Delta E_k(Q,z) = E_k(Q,z)-E_{k-1}(Q,z) \quad (k=1,\ldots,n)$$

are one dimensional orthogonal projections.

Write $Q(z) = zI - B(z)$ with

$$B(z) = z \int_0^{\eta} exp(-zs)d\tilde{R}_0(s) + \int_0^{\eta} exp(-zs)dR_0(s).$$

Then

$$Q^{-1}(z) = \frac{1}{z}(I - \frac{1}{z}B(z))^{-1}.$$

Now by (3.2) with $Q(z)$ instead of $I - A$, we get

$$Q^{-1}(z) = \frac{1}{z} \prod_{1 \le k \le n}^{\rightarrow} \left(I + \frac{B(z)\Delta E_k(Q,z)}{z - \lambda_k(B(z)}\right).$$

But

$$z - \lambda_k(B(z)) = \lambda_k(Q(z)).$$

Thus we arrive at our next result.

Theorem 9.4 *For any regular* $z \ne 0$ *of* $Q(.)$, *the equality*

$$Q^{-1}(z) = \frac{1}{z} \prod_{1 \le k \le n}^{\rightarrow} \left(I + \frac{B(z)\Delta E_k(Q,z)}{\lambda_k(Q(z))}\right)$$

is true.

9.4 Perturbations of Characteristic Values

Let

$$Q(z) = z(I - \int_0^{\eta} exp(-zs)d\tilde{R}_0(s)) - \int_0^{\eta} exp(-zs)dR_0(s) \tag{4.1a}$$

and

$$M(z) = z(I - \int_0^{\eta} exp(-zs)d\tilde{R}_1(s)) - \int_0^{\eta} exp(-zs)dR_1(s), \tag{4.1b}$$

where R_1 and $\tilde{R}_1$ are $n \times n$-matrix-valued functions defined on $[0, \eta]$, whose entries have bounded variations, R_0 and $\tilde{R}_0$ are the same as in Sect. 9.1.

Enumerate the characteristic values $z_k(Q)$ and $z_k(M)$ of Q and M, respectively with their multiplicities in the nondecreasing order of their moduli.

The aim of this section is to estimate the quantity

$$rv_Q(M) = \max_{j=1,2,\ldots} \min_{k=1,2,\ldots} \left| \frac{1}{z_k(Q)} - \frac{1}{z_j(M)} \right|$$

which will be called *the relative variation of the spectrum of* $M(.)$ *with respect to the spectrum of* $Q(.)$.

Assume that the matrices

$$B_0 = R_0(0) - R_0(\eta) \text{ and } T_0 = R_1(0) - R_1(\eta) \text{ are invertible}, \tag{4.2}$$

and

$$\eta < 1. \tag{4.3}$$

As it was shown in Sect. 9.1, the latter condition does not affect on the generality. Again put

$$Q_1(z) = B_0^{-1}(\eta)Q(z) \text{ and, in addition, } M_1(z) = T_0^{-1}M(z).$$

So

$$Q_1(z) = \sum_{k=1}^{\infty} \frac{z^k}{k!} C_k \text{ and } M_1(z) = \sum_{k=1}^{\infty} \frac{z^k}{k!} D_k, \tag{4.4}$$

where C_k are defined as in Sect. 9.1 and

$$D_0 = I,\, D_1 = T_0^{-1}\left(I - \int_0^{\eta} d\tilde{R}_1(s) + \int_0^{\eta} s\,dR_1(s)\right),$$

$$D_k = (-1)^k T_0^{-1}\left(k\int_0^{\eta} s^{k-1} d\tilde{R}_1(s) - \int_0^{\eta} s^k dR_1(s)\right) \quad (k \geq 2).$$

As it is shown in Sect. 9.1, condition (4.3) implies the inequalities

$$\sum_{k=1}^{\infty} \|C_k\|^2 < \infty \text{ and } \sum_{k=1}^{\infty} \|D_k\|^2 < \infty. \tag{4.5}$$

Recall that

$$\Psi_Q := \left[\sum_{k=1}^{\infty} C_k C_k^*\right]^{1/2}$$

and put

$$\gamma(Q) := 2N_2(\Psi_Q) + 2[n(\zeta(2) - 1)]^{1/2},$$

where $\zeta(.)$ is the Riemann zeta function, again. As it is shown in Sect. 9.1,

$$\|\Psi_Q\| \le \alpha(F_Q) := \|B_0^{-1}\| \left(1 + \frac{var(\tilde{R}_0)}{(1-\eta)^2} + \frac{var(R_0)}{1-\eta}\right)$$

and therefore

$$\gamma(Q) \le 2\sqrt{n}\,(\alpha(F_Q) + (\zeta(2) - 1)^{1/2}).$$

Denote also

$$\xi(Q, s) := \frac{1}{s} exp\left[\frac{1}{2} + \frac{\gamma^2(Q)}{2s^2}\right] \quad (s > 0) \tag{4.6}$$

and

$$q = \left[\sum_{k=1}^{\infty} \|D_k - C_k\|^2\right]^{1/2}.$$

Theorem 9.5 *Let conditions* (4.2) *and* (4.3) *hold. Then*

$$rv_Q(M) \le r(Q, q),$$

where $r(Q, q)$ is the unique positive root of the equation

$$q\xi(Q, s) = 1. \tag{4.7}$$

This result is a particular case of Theorem 12.5.1 from the book Gil' (2010a). If we substitute the equality $y = x\gamma(Q)$ into (4.7) and apply Lemma 1.6.4 from the book Gil' (2003), we get $r(Q, q) \le \delta(Q, q)$, where

$$\delta(Q, q) := \begin{cases} e\,q & \text{if } \gamma(Q) \le e\,q \\ \gamma(Q)\,[ln\,(\gamma(Q)/q)]^{-1/2} & \text{if } \gamma(Q) > e\,q \end{cases}.$$

Therefore, $rv_Q(M) \le \delta(Q, q)$.

Put

$$W_j = \left\{z \in \mathbf{C} : q\xi\left(Q, \left|\frac{1}{z} - \frac{1}{z_j(Q)}\right|\right) \ge 1\right\} \quad (j = 1, 2, \ldots).$$

Since $\xi(Q, y)$ is a monotone decreasing function with respect to $y > 0$, Theorem 9.5 yields the following result.

Corollary 9.4 *Under the hypothesis of Theorem 9.5, all the characteristic values of M lie in the set*

$$\cup_{j=1}^{\infty} W_j$$

Let us apply the Borel (Laplace) transform to estimate q. Namely, put

$$F_1(u) = \int_0^\infty e^{-ut}(M_1(t) - Q_1(t))\,dt \quad (u \in \mathbf{C};\ Re\ u > 0).$$

Then

$$F_1(u) = \sum_{k=1}^{\infty} \frac{1}{u^{k+1}}(D_k - C_k) \tag{4.8}$$

and

$$F_1(u) = -\int_0^\infty e^{-ut}\left[t\int_0^\eta exp(-ts)d(B_0^{-1}\tilde{R}_0(s) - T_0^{-1}\tilde{R}_1(s))\right.$$

$$\left.+\int_0^\eta exp(-ts)d(B_0^{-1}R_0(s) - T_0^{-1}R_1(s)\,)dt\right].$$

Therefore,

$$F_1(u) = -\int_0^\eta \frac{1}{(u+s)^2}d(B_0^{-1}\tilde{R}_0(s) - T_0^{-1}\tilde{R}_1(s)) + \int_0^\eta \frac{1}{u+s}d(B_0^{-1}R_0(s) - T_0^{-1}R_1(s)). \tag{4.9}$$

In view of (4.3) equalities (4.8) and (4.9) can be extended to the domain $|u| \geq 1$. Moreover,

$$\frac{1}{2\pi}\int_0^{2\pi} F_1^*(e^{-is})F_1(e^{is})ds = \sum_{k=1}^{\infty}(D_k^* - C_k^*)(D_k - C_k).$$

Thus,

$$\frac{1}{2\pi}\,Trace\int_0^{2\pi} F_1^*(e^{-is})F_1(e^{is})ds = Trace\sum_{k=1}^{\infty}(D_k^* - C_k^*)(D_k - C_k),$$

and consequently,

$$q^2 \leq \sum_{k=1}^{\infty} N_2^2(D_k - C_k) = \frac{1}{2\pi} \int_0^{2\pi} N_2^2(F_1(e^{is}))ds.$$

This forces

$$q^2 \leq \frac{1}{2\pi} \int_0^{2\pi} N_2^2(F_1(e^{is}))ds \leq \sup_{|z|=1} N_2^2(F_1(z)).$$

Since $N_2^2(F_1(z)) \leq n\|F_1(z)\|^2$, equality (4.9) implies $q \leq q_1$, where

$$q_1 = \int_0^{\eta} \frac{\sqrt{n}}{(1-\eta)^2} ||d(B_0^{-1}\tilde{R}_0(s) - T_0^{-1}\tilde{R}_1(s))||$$

$$+ \int_0^{\eta} \frac{\sqrt{n}}{1-\eta} \|d(B_0^{-1}R_0(s)) - T_0^{-1}R_1(s))\|.$$

9.5 Convex Functions of Characteristic Values

We need the following classical result.

Lemma 9.3 *Let $\phi(x)$ $(0 \leq x \leq \infty)$ be a convex continuous function, such that $\phi(0) = 0$, and*

$$a_j, b_j \quad (j = 1, 2, \ldots, l \leq \infty)$$

be two non-increasing sequences of real numbers, such that

$$\sum_{k=1}^{j} a_k \leq \sum_{k=1}^{j} b_k \quad (j = 1, 2, \ldots, l).$$

Then

$$\sum_{k=1}^{j} \phi(a_k) \leq \sum_{k=1}^{j} \phi(b_k) \quad (j = 1, 2, \ldots, l).$$

For the proof see for instance Lemma II.3.4 from the book Gohberg and Krein (1969), or Corollary IV.3.3 from the book Gohberg et al. (2000). Put

$$\chi_k = \omega_k(Q) + \frac{n}{k+n} \quad (k = 1, 2, \ldots).$$

Recall that $\omega_k(Q)$ are defined in Sect. 9.1.

The following result is due to the previous Lemma and Theorem 9.1.

Corollary 9.5 *Let $\phi(t)$ $(0 \le t < \infty)$ be a continuous convex scalar-valued function, such that $\phi(0) = 0$. Let conditions* (1.2) *and* (1.3) *hold. Then the inequalities*

$$\sum_{k=1}^{j} \phi\left(\frac{1}{|z_k(Q)|}\right) < \sum_{k=1}^{j} \phi(\chi_k) \quad (j = 1, 2, \ldots)$$

are valid. In particular, for any $p > 1$,

$$\sum_{k=1}^{j} \frac{1}{|z_k(Q)|^p} < \sum_{k=1}^{j} \chi_k^p$$

and thus

$$\left[\sum_{k=1}^{j} \frac{1}{|z_k(Q)|^p}\right]^{1/p} < \left[\sum_{k=1}^{j} \omega_k^p(Q)\right]^{1/p} + n\left[\sum_{k=1}^{j} \frac{1}{(k+n)^p}\right]^{1/p} \quad (j = 1, 2, \ldots). \tag{5.1}$$

For any $p > 1$ we have

$$\zeta_n(p) := \sum_{k=1}^{\infty} \frac{1}{(k+n)^p} < \infty.$$

Relation (5.1) with the notation

$$N_p(\Psi_Q) = \left[\sum_{k=1}^{n} \lambda_k^p(\Psi_Q)\right]^{1/p}$$

yields our next result.

Corollary 9.6 *Let the conditions* (1.2) *and* (1.3) *hold. Then*

$$\left(\sum_{k=1}^{\infty} \frac{1}{|z_k(Q)|^p}\right)^{1/p} < N_p(\Psi_Q) + n\zeta_n^{1/p}(p) \quad (p > 1).$$

The next result is also well known, cf. Lemma II.3.5 from the book Gohberg and Krein (1969), or Lemma IV.3.2 from the book Gohberg et al. (2000).

Lemma 9.4 *Let a scalar-valued function $\Phi(t_1, t_2, \ldots, t_j)$ with an integer j be defined on the domain*

$$0 < t_j \le t_{j-1} \ldots \le t_2 \le t_1 < \infty$$

and have continuous partial derivatives, satisfying the condition

$$\frac{\partial \Phi}{\partial t_1} > \frac{\partial \Phi}{\partial t_2} > \cdots > \frac{\partial \Phi}{\partial t_j} > 0 \text{ for } t_1 > t_2 > \cdots > t_j, \tag{5.2}$$

and $a_k, b_k (k = 1, 2, \ldots, j)$ *be two nonincreasing sequences of real numbers satisfying the condition*

$$\sum_{k=1}^{m} a_k \le \sum_{k=1}^{m} b_k \quad (m = 1, 2, \ldots, j).$$

Then $\Phi(a_1, \ldots, a_j) \le \Phi(b_1, \ldots, b_j)$.

This Lemma and Theorem 9.1 immediately imply our next result.

Corollary 9.7 *Under the hypothesis of Theorem* 9.1, *let condition* (5.2) *hold. Then*

$$\Phi\left(\frac{1}{|z_1(Q)|}, \frac{1}{|z_2(Q)|}, \ldots, \frac{1}{|z_j(Q)|}\right) < \Phi(\chi_1, \chi_2, \ldots, \chi_j).$$

In particular, let $\{d_k\}_{k=1}^{\infty}$ be a decreasing sequence of non-negative numbers. Take

$$\Phi(t_1, t_2, \ldots, t_j) = \sum_{k=1}^{j} d_k t_k.$$

Then the previous corollary yields the inequalities

$$\sum_{k=1}^{j} \frac{d_k}{|z_k(Q)|} < \sum_{k=1}^{j} \chi_k d_k = \sum_{k=1}^{j} d_k \left[\omega_k(Q) + \frac{n}{k+n}\right] \quad (j = 1, 2, \ldots).$$

9.6 Comments

The material of this chapter is adapted from Chap. 12 of the book Gil' (2010a).

References

Agarwal, R., Berezansky, L. Braverman, E. and Domoshnitsky A. (2012), *Nonoscillation Theory of Functional Differential Equations and Applications*, Elsevier, Amsterdam.

Agarwal, R. P. and Grace, S. R. (2000), Asymptotic stability of certain neutral differential equations, *Math. Comput. Model.*, *31*, 9–15.

Ahiezer, N. I. and Glazman, I. M. (1981), *Theory of Linear Operators in a Hilbert Space*. Pitman Advanced Publishing Program, Boston.

Aizerman, M.A. (1949), On a conjecture from absolute stability theory, *Ushekhi Matematicheskich Nauk*, *4 (4)*, 187–188. In Russian.

Ardjouni, A. and Djoudi, A. (2012), Fixed points and stability in neutral nonlinear differential equations with variable delays. *Opuscula Math.* *32*, no. 1, 519–529.

Azbelev, N.V., V.P. Maksimov and Rakhmatullina, L.F. (1995), *Introduction to the Theory of Functional Differential Equations*, Advanced Series in Math. Science and Engineering 3, Atlanta, GA: World Federation Publisher Company.

Azbelev, N.V. and Simonov P.M. (2003), *Stability of Differential Equations with Aftereffects*, Stability Control Theory Methods Appl. v. 20, Taylor & Francis, London.

Bainov D. and Domoshnitsky, A. (1993), Non-negativity of the Cauchy matrix and exponential stability of a neutral type system of functional-differential equations, *Extracta Mathematicae*, *8*, no. 1, 75–82.

Bazhenova, L.S. (2002), *The IO-stability of equations with operators causal with respect to a cone. Mosc. Univ. Math. Bull.* *57*, no.3, 33–35; translation from Vestn. Mosk. Univ., Ser. I, (2002), no.3, 54–57.

Bellen, A., Guglielmi, B. and Ruehli, A.E. (1999), Methods for linear systems of circuits delay differential equations of neutral type, *IEEE Trans. Circuits Syst.*, *46*, 212–216.

Berezansky, L. and Braverman, E. (2006a), On exponential stability of linear differential equations with several delays, *J. Math. Anal. Appl.* *324* 1336–1355.

Berezansky, L. and Braverman, E. (2006b), On stability of some linear and nonlinear delay differential equations, *J. Math. Anal. Appl.*, *314*, 391–411.

Berezansky, L. and Braverman, E. (2009), On exponential stability of a linear delay differential equation with an oscillating coefficient. *Appl. Math. Lett.* *22*, no. 12, 1833–1837.

Berezansky, L., Braverman, E. and Domoshnitsky, A. (2008) Stability of the second order delay differential equations with a damping term. *Differ. Equ. Dyn. Syst.* *16*, no. 3, 185–205.

Bhatia, R. (2007) *Perturbation Bounds for Matrix Eigenvalues*, Classics in Applied Mathematics, Vol. 53, SIAM, Philadelphia, PA.

M.I. Gil', *Stability of Neutral Functional Differential Equations*,
Atlantis Studies in Differential Equations 3, DOI 10.2991/978-94-6239-091-1

Bylov, B.F., Grobman, B.M., Nemyckii, V.V. and Vinograd, R.E. (1966), *The Theory of Lyapunov Exponents*, Nauka, Moscow (In Russian).

Cahlon, B. and Schmidt, D. (2011), Necessary conditions and algorithmic stability tests for certain higher odd order neutral delay differential equations. *Dynam. Systems Appl. 20*, no. 2–3, 223–245.

Chen, Y., Xue, A., Lu, R. and Zhou, S. (2008), On robustly exponential stability of uncertain neutral systems with time-varying delays and nonlinear perturbations, *Nonlinear Anal., 68*, 2464–2470.

Cheng, S.S. and Lin, Y.Z. (2005), Exact regions of stability for an investment plan with three parameters, *Appl. Math. E-Notes, 5*, 194–201.

Corduneanu, C. (2002), *Functional Equations with Causal Operators*, Taylor and Francis, London.

Daleckii, Yu. L. and Krein, M. G. (1971) *Stability of Solutions of Differential Equations in Banach Space*, Amer. Math. Soc., Providence, R. I.

Demidenko, G. V. (2009), Stability of solutions to linear differential equations of neutral type. *J. Anal. Appl. 7*, no. 3, 119–130.

Drici, Z., McRae, F.A. and Vasundhara Devi, J. (2005), Differential equations with causal operators in a Banach space. *Nonlinear Anal., Theory Methods Appl. 62*, no.2 (A), 301–313.

Döetsch, G. (1961), *Anleitung zum Praktishen Gebrauch der Laplace-transformation*. Oldenburg, Munchen.

Domoshnitsky, A. (1983), Extension of Sturm's theorem to equations with time-lag, *Differentsial'nye Uravnenija, 19*, 1475–1482. In Russian.

Drakhlin, M.E. (1986), Operator of the internal superposition in the space of summable functions, *Izv. VUZov, Mathematics*, no. 5, 18–23. In Russian.

Drakhlin, M.E. and Plyshevskaya, T.K. (1978), To the theory of functional differential equations, *Differentsial'nye Uravnenia, 14*, 1347–1361. In Russian.

Dunford, N. and Schwartz, J.T. (1966), *Linear Operators, part I*, Interscience Publishers Inc, New York.

Feintuch, A., Saeks, R. (1982) *System Theory. A Hilbert Space Approach*. Ac. Press, New York.

Fridman, E. (2001), New Lyapunov-Krasovskii functionals for stability of linear retarded and neutral type systems, *Systems Control Lett., 43*, 309–319.

Gantmakher, F. R. (1967) *Theory of Matrices*, Nauka, Moscow. In Russian.

Gel'fand, I.M. and Shilov, G.E. (1958), *Some Questions of Theory of Differential Equations*. Nauka, Moscow. In Russian.

Gel'fond A.O. (1967), *Calculations of Finite Differences*, Nauka, Moscow. In Russian.

Gil', M. I. (1983), On a class of absolutely stable systems, *Soviet Physics Doklady, 280(4)*, 811–815.

Gil', M.I. (1998), *Stability of Finite and Infinite Dimensional Systems*, Kluwer, N. Y.

Gil', M.I. (2000a), On bounded input-bounded output stability of nonlinear retarded systems, *Robust and Nonlinear Control, 10*, 1337–1344.

Gil', M.I. (2000b), On Aizerman-Myshkis problem for systems with delay. *Automatica, 36*, 1669–1673.

Gil', M. I. (2001), Existence and stability of periodic solutions of semilinear neutral type systems, *Dynamics Discrete and Continuous Systems, 7*, no. 4, 809–820.

Gil', M.I. (2002), Boundedness of solutions of nonlinear differential delay equations with positive Green functions and the Aizerman - Myshkis problem, *Nonlinear Analysis, TMA, 49*, 1065–1078.

Gil', M.I. (2003) *Operator Functions and Localization of Spectra*, Lecture Notes In Mathematics, vol. 1830, Springer-Verlag, Berlin.

Gil', M.I. (2008), Inequalities of the Carleman type for Neumann-Schatten operators, *Asian-European J. of Math., 1*, no. 2, 203–212.

Gil', M. I. (2009a) Absolute and input-to-state stabilities of nonautonomous systems with causal mappings, *Dynamic Systems and Applications 18*, 655–666.

Gil', M. I. (2009b), L^2-absolute and input-to-state stabilities of equations with nonlinear causal mappings. *Internat. J. Robust Nonlinear Control 19*, no. 2, 151–167.

Gil', M.I. (2010a) *Localization and Perturbation of Zeros of Entire Functions*. Lecture Notes in Pure and Applied Mathematics, 258. CRC Press, Boca Raton, FL.

Gil', M.I. (2010b), Perturbations of functions of diagonalizable matrices, *Electr. J. of Linear Algebra, 20*, 303–313.

Gil', M.I. (2010c) Norm estimates for functions of matrices with simple spectrum, *Rendiconti del Circolo Matematico di Palermo, 59*, 215–226.

Gil', M.I. (2011a) The L^p- version of the generalized Bohl - Perron principle for vector equations with delay, *Int. J. Dynamical Systems and Differential Equations, 3*, no. 4, 448–458.

Gil', M.I. (2011b), The Lp-version of the generalized Bohl-Perron principle for vector equations with infinite delay, *Advances in Dynamical Systems and Applications, 6*, no. 2, 177–184.

Gil', M.I. (2012a), Stability of vector functional differential equations: a survey, *Quaestiones Mathematicae*, 35, 1–49.

Gil', M.I. (2012b), Exponential stability of nonlinear neutral type systems, *Archives Control Sci., 22 (LVIII)*, no 2, 125–143.

Gil', M.I. (2012c), Estimates for fundamental solutions of neutral type functional differential equations, *Int. J. Dynamical Systems and Differential Equations, 4* no. 4, 255–273.

Gil', M.I. (2013a) *Stability of Vector Differential Delay Equations*, Birkhäuser Verlag, Basel.

Gil', M.I. (2013b), The generalized Bohl-Perron principle for the neutral type vector functional differential equations, *Mathematics of Control, Signals, and Systems (MCSS), 25(1)*, 133–145.

Gil', M.I. (2013c), On Aizerman's type problem for neutral type systems, *European Journal of Control 19*, no. 2, 113–117.

Gil', M.I. (2013d), Input-to-state stability of neutral type systems, *Discussiones Mathematicae, Differential Inclusions, Control and Optimization, 33 (1)*, 1–12.

Gil', M.I. (2013e), Estimates for functions of finite and infinite matrices. Perturbations of matrix functions, *International Journal of Mathematics, Game Theory and Algebra, 21*, no. 4/5.

Gil', M.I. (2014), A bound for condition numbers of matrices, *Electron. Journal of Linear Algebra, ELA, 27*, 162–171.

Gil', M.I. (2014a), The L^p-version of the generalized Bohl-Perron Principle for neutral type functional differential equations, *Differential Equ. Dynamical Systems*, published online 18 Febr. 2014.

Gil', M.I. and Cheng, S. S. (2007), Solution estimates for semilinear difference-delay equations with continuous time, *Discrete Dynamics in Nature and Society*, v. 2007, Article ID 82027, 8 pages.

Gohberg, I., Golberg, S. and Krupnik, N. (2000), *Traces and Determinants of Linear Operators*, Birkháuser-Verlag, Basel.

Gohberg, I. C. and Krein, M. G. (1969), *Introduction to the Theory of Linear Nonselfadjoint Operators*, Trans. of Math. Monographs, vol. 18, Amer. Math. Soc., Rhode Island.

Gusarenko, S.A. and Domoshnitsky, A. (1989), Asymptotic and oscillation properties of first order linear scalar functional-differential equations, *Differentsial'nye Uravnenija, 25*, no. 12, 2090–2103. In Russian.

Halanay, A. (1966) *Differential Equations: Stability, Oscillation, Time Lags*, Academic Press, New York.

Hale, J.K. (1977), *Theory of Functional Differential Equations*, Springer-Verlag, New York.

Hale, J. K. and Lunel, S. M. V. (1993), Introduction to Functional Differential Equations, Springer-Verlag, New-York.

Han, Q.-L. (2008), A new delay - dependent absolute stability criterion for a class of nonlinear neutral systems. *Automatica 44*, 272–277.

He, Y., Wu, M., She, J.-H., and Liu, G.-P. (2004), Delay-dependent robust stability criteria for uncertain neutral systems with mixed delays. *Systems and Control Letters, 51*, 57–65.

Hewitt, E. and Stromberg, K. (1969), *Real and Abstract Analysis*, Springer Verlag, Berlin.

Horn R.A. and Johnson, C.R. (1985), *Matrix Analysis*, Cambridge Univ. Press, Cambridge.

Istratescu, V.I. (1981), *Introduction to Liner Operator Theory*, Marcel Dekker Inc, New Yorl.

Izobov, N.A. (1974), Linear systems of ordinary differential equations. *Itogi Nauki i Tekhniki. Mat. Analis, 12*, 71–146, In Russian.

Kolmanovskii, V. and Myshkis, A. (1999), *Applied Theory of Functional Differential Equations*, Kluwer, Dordrecht.

Kolmanovskii, V.B. and Nosov, V.R. (1986), *Stability of Functional Differential Equations*, Ac. Press, London.

Korenevskii, D. G. and Kaizer, K. (1998), Coefficient conditions for the asymptotic stability of solutions of systems of linear difference equations with continuous time and delay, *Ukrainian Mathematical Journal, 50*, no. 4, 586–590.

Krasnosel'skii, M.A., Lifshits, J. and Sobolev A. (1989), *Positive Linear Systems. The Method of Positive Operators*, Heldermann Verlag, Berlin.

Krasnosel'skii, M. A. and Zabreiko, P. P. (1984), *Geometrical Methods of Nonlinear Analysis*. Springer-Verlag, Berlin.

Krein, S.G. (1971), *Linear Differential Equations in a Banach Space*, Transl. Mathem. Monogr, vol 29, Amer. Math. Soc.

Krein, S.G. (1972), *Functional Analysis*, Nauka, Moscow. In Russian.

Kurbatov, V. (1999), *Functional Differential Operators and Equations*, Kluwer Academic Publishers, Dordrecht.

Lakshmikantham, V., Leela, S., Drici, Z. and McRae, F. A. (2009), *Theory of Causal Differential Equations*, Atlantis Studies in Mathematics for Engineering and Science, 5. Atlantis Press, Paris.

Liberzon, M.R. (2006), Essays on the absolute stability, *Automation and Remote Control, 67*, no. 10, 1610–1644.

Lupulescu, V. (2004), Existence of solutions for nonconvex functional differential inclusions. *Electron. J. Differential Equations*, no. 141, 6 pp.

Lupulescu, V. (2008), Causal functional differential equations in Banach spaces. *Nonlinear Anal. 69* no. 12, 4787–4795.

MacCamy, R. C. and Wong, J. S. W. (1972), Stability theorems for some functional equations. *Trans. Amer. Math. Soc., 164*, 137–147.

Marcus, M. and Minc, H. (1964), *A Survey of Matrix Theory and Matrix Inequalities*. Allyn and Bacon, Boston.

Meyer-Nieberg, P. (1991), *Banach Lattices*, Springer - Verlag, New-York.

Michiels, W. and Niculescu, S.I. (2007), *Stability and Stabilization of Time-Delay Systems. An Eigenvalue-Based Approach*, SIAM, Philadelphia.

Mitrinović, D. S., Pecaric, J.E. and Fink, A.M. (1991), *Inequalities Involving Functions and their Integrals and Derivatives*, Kluwer Academic Publishers, Dordrecht.

Myshkis A. D. (1977), On some problems of theory of differential equations with deviation argument, *Uspechi Matemat. Nauk, 32(194)*, no 2, 173–202. In Russian.

Niculescu, S.I. (2001), *Delay Effects on Stability: a Robust Control Approach*, Lecture Notes in Control and Information Sciences, vol. 269, Springer, London.

Nam, P. T. and Phat, V. N. (2009), An improved stability criterion for a class of neutral differential equations. *Appl. Math. Lett., 22*, 31–35.

Ostrowski, A.M. (1952), Note on bounds for determinants with dominant principal diagonals, *Proc. of AMS, 3*, 26–30.

Ostrowski, A. M. (1973), *Solution of Equations in Euclidean and Banach spaces*. Academic Press, New York - London.

Park, J. H. and Won, S. (2000), Stability analysis for neutral delay-differential systems, *J. Franklin Inst., 337*, 67–75.

Partington J. and Bonnet, C. (2004), H^∞ and BIBO stabilization of delay systems of neutral type, *Systems & Control Letters 52*, 283–288.

Pepe, P. (2003), The Liapunov second method for continuous time difference equations, *Int. J. Robust Nonlinear Control, 13*, 1389–1405.

Pepe, P. and Verriest, E.I. (2003), On the stability of coupled delay differential and continuous time difference equations, *IEEE Trans. Autom. Control, 48*, no. 8, 1422–1427.

Pontryagin, L. S. (1955), On zeros of some elementary transcendental functions, *Trans. of AMS, Series 2, v. 1*, 96–110.

Popescu, D., Rasvan, V. and Stefan, R. (2004), Applications of stability criteria to time delay systems, *Electronic Journal of Qualitative Theory of Differential Equations*, Proc. 7th Coll. QTDE, no. 18, 1–20.

Rasvan, V. (1983), *Absolute Stability of Equations with Delay*, Nauka, Moscow. In Russian.

Rasvan, V. (2002), Delay independent and delay dependent Aizerman problem, in *Open Problem Book* (V. D. Blondel and A. Megretski eds.) pp. 102–107, 15th Int'l Symp. on Math. Theory Networks and Systems MTNS15, Univ. Notre Dame USA, August 12–16.

Rasvan, V. (2013), Discussion on: "On Aizerman's type problem for neutral type systems", *European Journal of Control 19*, no. 2 (2013), 117–122.

Sharkovsky A.N., Maistrenko Yu.L. and Romanenko E.Yu. (1993), *Difference Equations and their Applications. Mathematics and its Applications*. Kluwer Academic Publishers Group: Dordrecht.

Stewart, G.W. and Sun Ji-guang. (1990), *Matrix Perturbation Theory*. Academic Press, New York.

Sun, Y. and L. Wang, (2006), Note on asymptotic stability of a class of neutral differential equations, *Appl. Math. Lett., 19*, 949–953.

Varga, R. S. (2004), *Gerschgorin and his Cycles*, Springer Series in Computational Mathematics 36, Springer-Verlag, Berlin.

Vath, M. (2000), *Volterra and Integral Equations of Vector Functions*, Marcel Dekker.

Vidyasagar, M. (1993), *Nonlinear Systems Analysis*, second edition. Prentice-Hall, Englewood Cliffs, New Jersey.

Voronov, A. A. (1977), Systems with a differentiable nondecreasing nonlinearity that are absolutely stable in the Hurwitz angle. *Dokl. Akad. Nauk SSSR, 234*, no. 1, 38–41. In Russian.

Voronov, A. A. (1979), *Stability, Controllability, Observability*, Nauka Publ. House, Moscow. In Russian.

Vulikh, B. Z. (1967), *Introduction to the Theory of Partially Ordered Spaces*, Wolters-Noordhoff Scientific Publications LTD, Groningen.

Walther, H.-O. (2011), More on linearized stability for neutral equations with state-dependent delays. *Differ. Equ. Dyn. Syst. 19*, no. 4, 315–333.

Wang, X. and L. Liao, (2003) Asymptotic behavior of solutions of neutral differential equations with positive and negative coefficients, *J. Math. Anal. Appl. 279*, 326–338.

Wang, X., Shuyong Li and Daoyi Xu (2011), Globally exponential stability of periodic solutions for impulsive neutral-type neural networks with delays, *Nonlinear Dyn., 64*, 65–75.

Willems, J.C. (1971), The analysis of feedback systems, Res. Monograph No. 62, MIT Press, Cambridge, MA.

Wu, M., Y. He and She, J.-H. (2004), New delay-dependent stability criteria and stabilizing method for neutral systems, *IEEE Trans. Automat. Control, 49*, 2266–2271.

Xu, B. and Wang, Q. (2005), LMI approach for absolute stability of general neutral type Lurie indirect control systems. *J. Control Theory Appl. 3*, no. 4, 387–392.

Yakubovich V.A. and Starzhinskii, V.M. (1975), *Differential Equations with Periodic Coefficients*, John Wiley, New York.

Zhang, Z. and Wang, Z. (2001), Asymptotic behavior of solutions of neutral differential equations with positive and negative coefficients, *Ann. Differential Equations, 17 (3)* 295–305.

Index

M.I. Gil', *Stability of Neutral Functional Differential Equations*,
Atlantis Studies in Differential Equations 3, DOI 10.2991/978-94-6239-091-1

MIX
Papier aus verantwortungsvollen Quellen
Paper from responsible sources
FSC® C105338

If you have any concerns about our products,
you can contact us on
ProductSafety@springernature.com

In case Publisher is established outside the EU,
the EU authorized representative is:
Springer Nature Customer Service Center GmbH
Europaplatz 3, 69115 Heidelberg, Germany

Printed by Libri Plureos GmbH
in Hamburg, Germany